HUMBER COLLEGE
LAKESHORE CAMPUS
LEARNING RESOURCE CENTRE
3199 LAKESHORE BLVD. WEST
TORONTO, ONTARIO M8V 1K8

Consumption and the Transformation of Everyday Life

*Consumption and Public Life*

Series Editors: Frank Trentmann and Richard Wilk

*Titles include:*

Mark Bevir and Frank Trentmann (*editors*)
GOVERNANCE, CITIZENS AND CONSUMERS
Agency and Resistance in Contemporary Politics

Daniel Thomas Cook (*editor*)
LIVED EXPERIENCES OF PUBLIC CONSUMPTION
Encounters with Value *in* Marketplaces on Five Continents

Nick Couldry, Sonia Livingstone and Tim Markham
MEDIA CONSUMPTION AND PUBLIC ENGAGEMENT
Beyond the Presumption of Attention

Kate Soper and Frank Trentmann (*editors*)
CITIZENSHIP AND CONSUMPTION

Harold Wilhite
CONSUMPTION AND THE TRANSFORMATION OF EVERYDAY LIFE
A View from South India

*Forthcoming:*

Jaccqueline Botterill
CONSUMER CULTURE AND PERSONAL FINANCE
Money Goes to Market

Roberta Sassatelli
FITNESS CULTURE
Gyms and the Commercialisation of Discipline and Fun

---

Consumption and Public Life
Series Standing Order ISBN 1–4039–9983–X Hardback 1–4039–9984–8 Paperback
(*outside North America only*)

You can receive future titles in this series as they are published by placing a standing order. Please contact your bookseller or, in case of difficulty, write to us at the address below with your name and address, the title of the series and one of the ISBNs quoted above.

Customer Services Department, Macmillan Distribution Ltd, Houndmills, Basingstoke, Hampshire RG21 6XS, England

# Consumption and the Transformation of Everyday Life

## A View from South India

Harold Wilhite
*University of Oslo, Norway*

© Harold Wilhite 2008

All rights reserved. No reproduction, copy or transmission of this publication may be made without written permission.

No paragraph of this publication may be reproduced, copied or transmitted save with written permission or in accordance with the provisions of the Copyright, Designs and Patents Act 1988, or under the terms of any licence permitting limited copying issued by the Copyright Licensing Agency, 90 Tottenham Court Road, London W1T 4LP.

Any person who does any unauthorized act in relation to this publication may be liable to criminal prosecution and civil claims for damages.

The author has asserted his right to be identified as the author of this work in accordance with the Copyright, Designs and Patents Act 1988.

First published 2008 by
PALGRAVE MACMILLAN
Houndmills, Basingstoke, Hampshire RG21 6XS and
175 Fifth Avenue, New York, N.Y. 10010
Companies and representatives throughout the world

PALGRAVE MACMILLAN is the global academic imprint of the Palgrave Macmillan division of St. Martin's Press, LLC and of Palgrave Macmillan Ltd. Macmillan® is a registered trademark in the United States, United Kingdom and other countries. Palgrave is a registered trademark in the European Union and other countries.

ISBN 13: 978–0–230–54254–9   hardback
ISBN 10: 0–230–54254–9   hardback

This book is printed on paper suitable for recycling and made from fully managed and sustained forest sources. Logging, pulping and manufacturing processes are expected to conform to the environmental regulations of the country of origin.

A catalogue record for this book is available from the British Library.

Library of Congress Cataloging-in-Publication Data

Wilhite, Harold, 1946–
  Consumption and the transformation of everyday life : a view from South India / Harold Wilhite.
    p. cm. – (Consumption and public life)
  Includes bibliographical references and index.
  ISBN 0–230–54254–9 (alk. paper)
    1. Consumption (Economics) – India.  2. India – Economic conditions.
  3. India – Social conditions.  I. Title.

HC440.C6W55 2008
339.4'709548–dc22                                           2008015909

10  9  8  7  6  5  4  3  2  1
17 16 15 14 13 12 11 10 09 08

Printed and bound in Great Britain by
CPI Antony Rowe, Chippenham and Eastbourne

# Contents

| | |
|---|---|
| *List of Maps* | viii |
| *List of Figures* | ix |
| *List of Tables* | x |
| *List of Plates* | xi |
| *List of Abbreviations* | xii |
| *Acknowledgements* | xiii |
| *Malayalam Glossary* | xiv |

**1 Introduction** — 1
- What is consumption? — 3
- Debates about individual, social and material contributions to consumption — 5
- Debates about global and local — 7
- The vocabulary of modern and traditional — 8
- The middle class — 9
- The study — 10
- Organisation of the book — 13

**2 Global Interchange and Modernising Reforms** — 15
- A history of global contact and migration — 17
- Kerala socialism — 19
- Reforms in family and gender relations — 20
- Caste reforms and the blurring of caste hierarchy — 22
- Conclusion — 29

**3 Women in a Bind: The Crucible of Marriage and Dowry** — 31
- Social scrutiny — 33
- Mobility — 34
- Dress — 35
- Beauty — 38
- From background to foreground — 40
- Beauty from elsewhere — 42
- Constraints, ambiguities and changes in small steps — 48

| | |
|---|---:|
| **4 The Modern Housewife** | **49** |
| A day in the life of Chavita and Anuj | 49 |
| Gender sharing of tasks in the home | 53 |
| The sources of gender ideology | 56 |
| Clothes washing | 60 |
| Food preparation | 61 |
| Conclusion | 65 |
| **5 Exercising the Extended Family** | **68** |
| Events and celebrations that reinforce family | 73 |
| Meena's first pregnancy | 77 |
| How dowry is implicated in household economy and consumption | 78 |
| Meena's dowry | 80 |
|    The constituents of dowry | 84 |
|    The embedded cost of education | 86 |
|    Marshalling dowry | 86 |
| Conclusion | 88 |
| **6 Work Migration** | **89** |
| Cavita's migration story | 91 |
| People from nowhere | 94 |
| Consumption in 'Gulf' families | 98 |
| The pull of family | 99 |
| Dual residence | 101 |
| Conclusion | 103 |
| **7 Material, Discursive and Performative Contributions to Consumption** | **104** |
| Keeping clean | 104 |
| Hot water consumption | 106 |
| Soap as development | 107 |
| Keeping cool | 111 |
|    Changes in the design and materials in home construction | 113 |
|    Changes in the political economy of air conditioners | 117 |
|    Alternatives to air conditioning | 118 |
|    The normalising of air conditioning | 119 |
| Modernity on wheels | 121 |
| Conclusion | 126 |
| **8 Frictionless Political and Religious Ideologies** | **128** |
| Gandhi's legacy | 128 |
| The fall and rise of *swadeshi* | 130 |

|   | Kerala socialism | 136 |
|---|---|---|
|   | Religiosity | 140 |
|   |   Pilgrimages | 141 |
|   |   Devotionalism | 141 |
|   |   Christian religiosity | 144 |
|   | The decline of frugality | 146 |
| **9** | **Television: 'Everyone is Watching'** | **148** |
|   | The advent of television in India and Kerala | 150 |
|   | An evening at the serials | 152 |
|   | *Sthree Malayalam* (Malayalee women) | 152 |
|   | *Sthree oru jwala* (Woman, Aflame) | 155 |
|   | *Sthreejanmam* (Women's Lives) | 156 |
|   | Television advertising | 158 |
|   | Conclusion | 163 |
| **10** | **Conclusion** | **165** |
|   | Transnational agents | 165 |
|   | Social relations and cultural practices | 168 |
|   | Technology | 170 |
|   | Consumption as performance | 172 |
|   | Consumption and the environment | 173 |
| *Notes* | | 177 |
| *References* | | 185 |
| *Index* | | 198 |

# List of Maps

Map 1  Map of India, with Kerala and its capital  xvii
       Thiruvananthapuram (Trivandrum)
Map 2  Trivandrum  11

# List of Figures

| | | |
|---|---|---|
| Figure 2.1 | Distribution of Hindu Nair and Ezhava households in four middle class Trivandrum neighbourhoods, based on a survey of 408 households | 26 |
| Figure 2.2 | Percentage of female heads of households of Ezhava and Nair castes having completed secondary school, bachelors and masters degrees, based on a survey of 408 households in four Trivandrum neighbourhoods | 27 |
| Figure 2.3 | Ownership of household appliances, cars and motorcycles among Hindu Ezhava and Hindu Nair families, based on a survey of 245 households in four Trivandrum neighbourhoods | 28 |
| Figure 3.1 | Percentage of women using selected beauty or cosmetic products more than once a week, based on a survey of 408 households in four Trivandrum neighbourhoods | 40 |
| Figure 4.1 | Differences in ownership of convenience appliances between households with female heads of household working outside the home and households with female heads of household not working outside the home (based on a survey of 408 households in four neighbourhoods in Trivandrum) | 56 |
| Figure 6.1 | Ownership of selected appliances comparing families with a family member working outside India (overseas) versus families with no family member working outside India (no overs), based on a survey of 408 households in four Trivandrum neighbourhoods | 99 |
| Figure 6.2 | Ownership of selected appliances comparing families with a member working outside of India (overseas) versus families with no family member working outside (no overs), by household income categories, based on a survey of 408 households in four Trivandrum neighbourhoods | 100 |

# List of Tables

| | | |
|---|---|---|
| Table 4.1 | Participation of male and female heads of households in selected home chores and activities, based on a survey of 408 households in four Trivandrum neighbourhoods | 54 |
| Table 4.2 | Amount of time used daily on selected household chores and activities by females and males in the household, based on five days of diaries kept by all adult members of 20 households | 54 |
| Table 4.3 | Participation of male and female heads of households in selected home chores and activities in households in which women work full time outside the home, based on a survey of 408 households in four Trivandrum neighbourhoods | 55 |
| Table 5.1 | Average cash, gold and land transferred in dowry, based on a survey of 408 households in four Trivandrum neighbourhoods | 84 |
| Table 5.2 | Average cash, gold and property components of dowry transferred according to the monthly income in households, based on a survey of 408 households in four Trivandrum neighbourhoods | 87 |

# List of Plates

| | | |
|---|---|---|
| Plate 1 | My Fair Lady Beauty Parlour, Trivandrum | 39 |
| Plate 2 | University students | 45 |
| Plate 3 | Syrian Christian married woman in *churidar* and *kurta* | 46 |
| Plate 4 | Hindu Nair family, dressed for public. Man in shirt and pants, woman in *sari* | 50 |
| Plate 5 | Hindu Ezhava family, dressed for home. Man in shirt and *lungi*, woman in 'house dress' | 59 |
| Plate 6 | The *chorunu* (rice feeding) ceremony | 74 |
| Plate 7 | The *Vishukanni* | 75 |
| Plate 8 | AC advertisement, Kumarapuram, Trivandrum | 112 |
| Plate 9 | Gulf house, Kawdiar, Trivandrum | 115 |

# List of Abbreviations

| | |
|---|---|
| CDS | Centre for Development Studies |
| OBC | Other backward castes |
| UAE | United Arab Emirates |
| NRI | Non Resident Indian |
| SUV | Sports Utility Vehicle |
| TNC | Transnational Corporations |
| CPI | Communist Party of India |
| IMF | International Monetary Fund |
| BJP | Bharatiya Janata Party |
| KSSP | Kerala Sasatra Sahita Parishat |
| CI | Consumer's International |
| SST | Social Shaping of Technology |
| WCDE | World Commission on the Environment |

# Acknowledgements

This book would not have been realised without the support and assistance of my three Malayalee guides, Alwin Jayakumar, Renu Henry and Ajith Kumar. Their support and their kindness will never be forgotten. I also want to acknowledge the patience of neighbours in Trivandrum and the interview participants who generously shared their time with me. I am also indebted to the Energy Management Centre, Government of Kerala, for its cooperation, as well as to J. Devika and Mridul Eapen at the Centre for Development Studies (CDS) in Trivandrum for their insights and suggestions of literature and contacts. I thank Aleyamma Vijayan of *SAKHI*, Sister Patricia of *PRANA* and C. S. Soman of *Health Action by People* for many fruitful conversations and for their help with the survey questionnaire.

At the Centre for Development and the Environment (SUM), University of Oslo, I want to express appreciation to Desmond McNeill, for working actively with me in getting financial support for the project. I am grateful to the Norwegian Research Council for funding the project. At the Department of Social Anthropology, Marianne Lien, read and commented several versions of the manuscript. I want to also thank Fillipo Osella (Sussex University), Michael Selzer (Oslo City University), Elizabeth Shove (Lancaster University), Daniel Miller (University College London) as well as Tanja Winther, Katinka Frøystad and Alida Boye here at the University of Oslo. Financial support for the research was provided by the Norwegian Research Council and its Programme entitled 'Programme for Sustainable Consumption and Production'. Thanks to the Programme Secretary, Knut Bjørseth for his encouragement and backing.

Finally, I could not have undertaken this project without the support of my wife Zehlia and our children, Paul, Alexandre and Kahena. Zehlia not only encouraged me, but was an active participant in the field work in Kerala. She opened pathways into life in Trivandrum which would have been difficult, if not impossible to pursue without her.

# Malayalam Glossary

*Achappam* (or *appam*)   Rice flour pancakes mixed with sugar and coconut milk.
*Ahimsa*   The Gandhian principles of truth and non-violence.
*Attukal Pongala*   Important temple festival in Trivandrum, attended by women.
*Ayurvedic*   A form of medical practice or medicine based on herbal products.
*Bindi*   A cosmetic mark on the forehead. Hindu married women mark their foreheads in red, but bindi is increasingly being used by unmarried women and Christian women.
*Brumana*   The spiritual haven achieved by Hindus after release from the cycle of physical reincarnation.
*Chai*   A drink composed of tea, milk, sugar and spices, served hot.
*Chara*   Free roving.
*Chorunu*   The feeding ceremony for Hindu children, the first time they are given food to supplement mother's milk.
*Chula*   Cooking oven which is fueled by a fire. Coconut rinds and other forms for biomass are burned.
*Churidar*   A garment worn by women, mainly young unmarried women, as daily wear. Less formal than the *sari*.
*Dhoti*   A long loincloth worn by Hindu men.
*Dobi*   The caste which has traditionally washed the clothes of upper caste.
*Ezhava*   Caste formerly defined as lower caste, traditionally related to work with coconut trees.
*Gram Sabhas*   Political forum at the local community level.
*Idyli*   Rice cakes.
*Irumudi*   Offerings for god Ayyappan.
*Karnavar*   Senior male head of *taravad*, with continued usage for male head of patrilineal family.
*Lakh*   100,000.
*Lungi*   A garment worn by men, a cotton cloth folded and tied around the waste, resembling a skirt.
*Mahabharata*   An important Hindu saga.
*Mahout*   Elephant caretaker.
*Makadam*   January.
*Marumakkathayam*   Matrilineal system of kinship and inheritance.
*Moksha*   In Hinduism, spiritual salvation that gives freedom from the cycle of death and rebirth.
*Munde*   A garment worn by men for more formal occasions than the *lungi*, a cotton cloth folded around the waste resembling a skirt.
*Murithirikothu*   A snack made of green ground powder, rice powder, and jaggery.
*Muryukka*   A snack made from black gram powder, fried in oil.
*Nair* (also *Nayar*)   Caste considered as upper caste, landowners, administrators and teachers.
*Nambuthiri Brahmin*   Caste earlier considered the highest caste, priests and landowners. Today hold many kinds of jobs.

## Malayalam Glossary xv

*Nayapam*   A dish made of rice, jaggery and cardamom, all fried in oil.
*Pakkavada*   A spicy mix of fried chilli, gram powder and rice powder.
*Onam*   Important festival marking the anniversary of the mythical founding of Kerala.
*Panchayat*   An administrative unit of local government roughly equivalent to county.
*Pappadam*   A thin pancake made of flour and fried crisp before eating.
*Pariah*   A former untouchable caste that worked in crematoriums and slaughter houses.
*Parasudi*   Very clean or pure.
*Pativrata*   Hindu concept of chastity.
*Payasam*   A dessert prepared with rice or other grains, along with milk and sugar or molasses.
*Perideel*   The naming ceremony for Hindu children.
*Pranapara*   Breathing technique.
*Pudava*   Clothes given to a woman on her marriage.
*Pula*   Ritual pollution.
*Pulaya*   A former untouchable caste that did rough agricultural work.
*Puja*   A form for worship associated with spiritual cleansing.
*Ramayana*   An important Hindu saga.
*Sabarimala*   Annual pilgrimage to visit the shrine of the god Ayyappan, open for men and women of non-menstruating ages.
*Sadhu*   A Hindu holy man.
*Sambandham*   Marriage-like liaisons under the matrilineal system of kinship.
*Sambar*   A curry dish prepared with vegetables, dhal and tamarind.
*Sanjaanam*   The daily lighting of a lamp or candle in honor of ancestors.
*Sanyas*   The s*anscrit* word for renunciation.
*Sari*   A garment worn by women outside the house.
*Sudhi*   Clean.
*Swabhaavam*   Socially accepted behaviour for women.
*Stridhanam*   Bride's wealth (dowry) under the matrilineal kinship system.
*Swadeshi*   Local, or indigenous development.
*Swaraj*   Self reliance.
*Thachusasthram*   Principles for house design and construction.
*Taravad*   House and landholdings of Nair matrilineal families.
*Talikettukalyanam*   The ceremony during which men and women were joined together in *sambandham*.
*Thaapabharani*   A thermally insulated box for completing the cooking of rice after it begins to simmer.
*Thali*   Necklace which is used as a marriage symbol.
*Toddy*   Palm wine.
*Upabhogam*   Consumer.
*Upanishads*   Hindu writings on metaphysics.
*Vedas*   Hindu books of knowledge.
*Vethideel*   Herbal bath.
*Viswakamos*   Artisan caste, traditionally responsible for building houses for upper caste families.
*Vishu*   The celebration in April, when the Hindu calendar places the earth in an auspicious position in relation to the sun.

*Vishukkani* An elaborate display of fruit, jewels, coins, flowers and a burning lamp, prepared by each family for their first sight after waking on the morning of *Vishu*.

*Wyakkuaruy* Rice placed in the mouth of a deceased person by relatives in order to provide sustenance for the journey into the next life.

**Map 1** Map of India, with Kerala and its capital Thiruvananthapuram (Trivandrum).

# 1
# Introduction

This book takes on the question of why the consumption of household goods and commodities are growing and changing rapidly in India. The book is based on a study centred in the state of Kerala in southern India, but the questions raised in the book about consumption and social change have relevance for India, Asia and many other parts of the world. Whether measured in terms of material objects, the raw materials embedded in them, or the energy and water that make it possible to use them, household consumption continues to grow in the rich OECD countries and has grown rapidly in many places in the 'South' over the past two decades (OECD 2004). On the surface, this growth in consumption can simply be attributed to growth in national economies and a more integrated global economy. However, there is much evidence that consumption growth is not simply a function of economic growth. The state of Kerala is a good example. Measured in terms of overall consumer spending, consumption in Kerala is higher than in any other Indian State (National Sample Survey Organisation 2001/2). Consumption of household durable goods (including household electrical appliances and vehicles) in Kerala is four times the national average. This high consumption relative to other Indian states is taking place in a state with only moderate economic growth. Growth in Kerala ranks near the bottom compared with other Indian states (Surendran 1999; OECD 2004; CI-ROAP 1998).

Kerala's high consumption and low economic growth relative to the rest of India raises intriguing questions about the reasons behind increasing consumption and is one good reason for locating a study of consumption in Kerala. Another is Kerala's political heritage, with its emphasis on public education, health and a publicly planned economy. India's highest consuming state is also one with an anti-capitalist political agenda.[1] Kerala consumption is also interesting because it is related to work migration, which is extensive in Kerala, and is also growing in other parts of India and Asia. It is estimated that repatriated income from Kerala's work migrants accounts for as much as 40 per cent of Kerala's GDP (Zachariah et al. 2002b). However, little is understood about the relationship between migration and consumption. Another important relationship is that of global capitalism and local consumption. From the early 1990s India has experienced a rapid entry into the global economy and a consequent rush into the country of capital, products and foreign media. In consumption practices ranging from the creation of personal appearance, to refrigeration of foods, to the building of a house, local consumption practices are changing in the meeting of local practices with several important facets of globalisation, including work migration and economic liberalisation. The meeting of local and global political economies, ideas and practices in Kerala, and the resulting changes in consumption make it an ideal place to study consumption practices and why they are changing.

This study joins a number of other recent efforts to bring ethnographic approaches to the study of changing consumption in the South. The work of Miller (1994) in Trinidad, Wilk (1996, 2002a) in Belize, Freeman (2000) in Barbados, Liechty (2003) in Nepal, Colloredo-Mansfeld (1999) in Ecuador, Hansen (2000) in Zambia, Mazzarella (2003) in India, Johnson (1998) in the Philippines, and O'Dougherty (2002) in Brazil are examples of important work that form cross-cultural references for changes in Kerala. Virtually all of these studies address the interface between globalising and local political economies; the interplay of modernity versus tradition; and the impacts of work migration. Contrasted with these studies, what is unique in this study centred on Kerala is the attention placed on the consumption of heavy durable goods, such as household electrical appliances and the house itself.

Bourdieu's practice theory has been useful in contextualising consumption in relation to other socio-cultural practices (Bourdieu 1977). A practice-theory approach is highly compatible with the ethnographic research method, in which frameworks for interpretation emerge from the study of practice in a given place. As Frøystad (2005:277) expressed theory-practice relationship: 'To a large extent the analytical perspectives that came to underpin this study (her study examined "Hinduness" in a northern Indian city) grew out of the material itself...While searching for patterns in my observations, I searched for theoretical frameworks that could help me to interpret and discuss them.'

Frøystad's search for theoretical frameworks in plural is important in a study of consumption. There is no single, composite theory that works for all of the types of consumption in the home. A search for an understanding of consumption practices as various as cleanliness, mobility and thermal comfort requires openness to the relevance of differing explanatory frameworks. As Moore (1999:4) argued, for subjects like consumption, 'It is no longer possible to speak of coherent theoretical approaches that are neatly delineated from others. Theories are themselves more composite, more partial and more eclectic.' As I engaged with consumption in Kerala's capital city of Trivandrum[2] several theoretical perspectives emerged as cogent, including Miller (1994; 1995b; 1995c; 2001a), Appadurai (1986; 1996) and Wilk (1996; 2001) on globalisation; Mankekar (1999), Liechty (2003) and Mazzarella (2003) on media, marketing and advertising; and Shove (2003) and Warde (1996), with their focus on the 'inconspicuous' consumption of unglamorous household goods. Another useful theoretical perspective is one that explores the power, or agency in consumption of technologies-in-use. The work of Latour (2000) and Akrich (2000) provide insights on how complex household appliance technologies (refrigerators, washing machines, air conditioners, and so on), once in place in the home, affect consumption practices in sometimes unforeseen ways.

## What is consumption?

In this study, consumption is conceptualised as the acquisition and use of things, including goods, products and increasingly, household appliance technologies. The essential questions in an analysis of consumption are first, how do products find their ways into people's homes; and second, how do they affect and get affected by daily practices. This conceptualisation moves the focus from the retail purchase to the social and geographical pathways from purchase to user. As we will see, in Kerala the retail purchase is often simply the beginning or end of a complex journey (or cultural biography in the words of Kopytoff 1986) which is implicated in, and influenced by social contexts such as those of family, marriage and migration. Household economics and retail decision-making are important to consumption decisions, but fall short of capturing the most important reasons behind changing consumption.

To reiterate, consumption involves acquisition and use. However, it is important to distinguish the acquisition of things intended for use by the consumer and those things intended to be used for production, such as acquiring a washing machine for the purpose of setting up a home-based laundry service. Consumption is also distinguished from savings and investments, where the intention is to generate capital or income from interest (Crocker and Linden 1998). Finally, consumption of many kinds of new household technologies brings with it the consumption of energy, water and other kinds of supporting products. Turning things around, the consumption of home energy, water and soaps is embedded in the consumption of the household appliances and cars. I have argued elsewhere that this point is not sufficiently acknowledged in the theory and policies of sustainable energy consumption (Wilhite 2005; Wilhite *et al.* 2000). The consumption of household appliances is one important reason for the rapid increase in electricity consumption in Kerala over the past decade. Residential electricity demand nearly doubled in Trivandrum from 1995 to 1999 and household electricity consumption grew from 16 per cent to 44 per cent of total household consumption over that same period (Vijayakumar and Chattopadhyay 1999). This rapidly increasing residential demand for electricity has contributed to severe shortages. Scheduled blackouts are a daily occurrence, and unforeseen blackouts usually happen several times daily. Augmenting the electricity production capacity to meet increased demand is expensive and environmentally problematic. There have been major political confrontations in recent years around the building of hydro-power plants because they necessitate the displacement of people in this densely populated state. In fact all of the economically feasible sources for producing new electricity are environmentally problematic. The least expensive of these, the coal-burning thermal power plant, contributes to several forms of serious local air and water pollution, as well as to the production of the climate-change gas carbon dioxide. Kerala's problem is a familiar one worldwide – meeting its growing demand for electricity will require the construction of new power sources, but the most economically feasable of these are environmentally problematic.

Consumption is also linked to the declining access to clean water. New washing practices involve an increasing use of soaps, both for cleaning bodies, clothes and homes. Soap use in turn brings with it increased water consumption for cleaning and rinsing. Most household waste runs untreated into water sources in Kerala and elsewhere in India.[3] Since agricultural runoff and industrial wastes in Kerala are also largely untreated, there has been a rapid decline of access to clean water (Argarwal, Narain and Khurana 2001). In many parts of Kerala, drinking water must be boiled or filtered, adding to chores performed mainly by women. Boiling water puts strains on household economies because it requires the use of either a commercial fuel such as bottled gas or kerosene, or biomass in the form of coconut rinds, which take time to gather. The shortage of clean water is an important factor behind the rapidly increasing consumption of bottled water, which is expensive and which contributes to another serious environmental problem in the form of plastic waste.

In short, another very good reason for examining changing consumption in Kerala and other places where the middle classes are growing is the environmental impacts of the increasing consumption of household appliances, cars, chemical cleaning solvents, soaps. The environmental consequences of consumption are addressed in this book, but the emphasis in this study has not been about what happens at the 'end of the pipe' – in the vernacular of environmental research – where the pollution runs out into the waterways and up into the atmosphere, but rather back at the 'beginning of pipe', in homes where the consumption is taking place. The intention is to contribute to a more robust understanding of consumption and of how it is changing, something that is direly needed in the research and policy worlds concerned with promoting environmental sustainability.

## Debates about individual, social and material contributions to consumption

On of the unique aspects of this study of consumption is the attention given to social relations and the material constructs of everyday life. This goes against the mainstream in consumption studies, which has been dominated by economic theory and method. Slater, who reviewed social science approaches to consumption, wrote that economics is 'the general model of social order through which the consumer is defined' (1997:41). In economic analyses, consumption is conceived as transactions made by economically rational individuals operating in a social world which is reduced to market or retail interactions. Consumption is stripped from everyday practice and actors from their social interactions. Slater is one of a long line of anthropologists and other social scientists who have critiqued these reductionist assumptions, extending back to Polanyi (1957) (one of the founders of economic anthropology). This is not to say that price, income and other economic considerations are not important to consumption; in Kerala, middle class access to capital and income are one part of the explanation for growing consumption. However, the reasons why people have used significant amounts of their income on consumption, and why certain kinds of consumption have been favoured over others are not accessible through a theoretical lens that makes oversimplifying assumptions about social practices, material infrastructures and discourses of development and media. As Hansen (2000:13) puts it, 'consumption (in the conventional economic view) is seen as the end point of the economic circuit' and 'focus on commodities rather than on consumers and leaves little scope to explore the workings of consumption and the diverse and changing social relations that individuals and collectivities are constructing through their consumption of objects'. The ambition in this book is to situate the study of consumption in everyday practices and to explore the social, material and discursive contributions to changing consumption.

In some ways, the approach in this book recalls some of the early work on the role of social structures and relations in human agency (Bourdieu 1977; Giddens 1979; Slater 1997; Lawson 1997). In much of the work of the 1980s and 1990s, the role in consumption of conventions, norms, social obligations and relations of power was ignored (Baumann 1988; Brown and Turley 1997; Lash and Urry 1994). As Warde wrote about that period 'A decade or more of analysis, founded in political economy and developing a materialist perspective on social life, seemed suddenly to be abandoned for a mode of studying culture which operated on wholly antithetical assumptions, according signs, discourses and mental constructs an exclusive role in understanding social activity (1997:1).' The 'materialist perspective' has been important in this study of consumption in Kerala, as has been the practice theory perspective of Bourdieu, with his emphasis on the importance of 'structured dispositions' in human agency and for the necessity to incorporate the actors 'practice, action, interaction, activity, experience, and performance' when theorising acts like those involved in consumption (1997:3). Pinches succinctly captures the theoretical perspective in this book, which views consumption as implicated in the 'material world of social experience (1996:5)'.

In short, social and material perspectives have been essential to understanding consumption and change in Kerala. Concerning clothes washing and the consumption of washing machines, a perspective on gender relations and the social organisation of work is absolutely essential to understanding change. Dramatic changes in food practices in the course of a few generations are only accessible through a perspective that incorporates the agency embedded in refrigeration technologies and the interaction between the technology and culturally-grounded ideas about health. An analysis of the consumption of air conditioning shows the importance of historical changes in the built environment as well as changes in the social organisation of work. The importance of a socio-material perspective to changes in cleanliness, comfort and food will be explored in coming chapters, as well as to subjects as diverse as beauty and mobility.

## Debates about global and local

The rapidly globalising economy has brought about important debates about consumption and socio-cultural change in countries of the South. World Systems Theory proposed that the force of global capitalism inevitably wipes over local culture and replaces it with Western values (Wallerstein 1974). This was also a basic premise of dependency theory (Cardoso and Faletto 1979; Frank 1972) and several of the more recent anti-globalisation theorists (see Falk 1999). In India, wariness about Western values and the power of expanding Western consumption was an important element in Gandhi's political ideals. This influenced the new Indian state's politics regarding transnational agents and foreign products in the decades following Independence in 1947. However, this relationship to global markets and expanding capitalism was to change dramatically in 1991 with the 'opening' of India and the Indian economy. In Chapter 8, the contribution to changing consumption of India's 'opened' economy is explored.

Miller (1994 and 1995a) and Wilk (1989) argued that the homogenising view of global capitalism is an oversimplification. Based on their ethnographies of cultural reproduction in Trinidad and Belize, respectively, they claimed that the products of global markets have the potential to invigorate rather than wipe over local culture. As Miller wrote, the entry of countries into cash economies and capitalist system 'may be as much the foundation for new diversities as for the elimination of old ones (1995a: 286)'. Miller (2001a:4) later moderated this view somewhat, writing that:

> At that time (in Marxist and World Systems theories) there was little sympathy for any approach to consumption that refused to regard it as other than merely the final outcome and a symptom of capitalism. By contrast, today the problem is to rescue consumption from being seen as a mode for the free expression of the creative subject.

Wilk (2001:52) calls for a theoretical middle ground between homogenisation and empowerment:

> The challenge for a theory of consumption in the third world is not to choose one bias or the other (coercion or autonomy, hegemony

or resistance). The task is instead to map out the areas where greater autonomy exists, and those where coercion takes place, and to seek an explanation for the variation in these areas...

In this study in south India, I have taken up Wilk's challenge, mapping how global forces and local socio-cultural contexts interact, though I will argue that global and local in India are far from neat, self-contained categories (see Tsing 2002). Chapter 2 traces Kerala's long history of global interaction. Two historically recent forms for globalisation, the 'opening' to global capitalism and work migration have had important effects on consumption in Kerala. The former has brought transnational enterprises, new product categories, as well as new forms for marketing and advertising, all backed by huge flows of money. Work migration has also channelled money and goods into Kerala. In scholarly analyses of consumption, work migration is often facilely attributed as the sole explanation of Kerala's high consumption relative to other States (Zachariah *et al.* 2002a); however, thus far none of these studies have examined how work migrants consume or how goods and ideas from places of work outside India get normalised in Kerala. Neither the effects on consumption of global capitalism or work migration can be understood without an accounting of the ways they engage with Kerala's extended family, cultural practices and local ideas about gender, religion and politics. These insights are relevant not only for Kerala, but for India as a whole and the many other countries of the South in which work migration is extensive and growing (see Gamburd 2000; Gardner 1995; and Johnson 1998).

## The vocabulary of modern and traditional

At the outset of the study, my intention was to avoid the vocabulary of modern and traditional. The meanings of these two terms are diffuse and have pejorative connotations, traditional being associated with backwardness and modern with progress. In spite of my reluctance to engage with modern and traditional, I was to find that these concepts were important in the vocabularies of Malayalee,[4] as well as in many sources of public discourse, such as government publications, in the local press and in the works of local scholars. Being or becoming 'modern' has been at the heart of social and political reforms in Kerala directed at the elimination of matriliny, reforming caste, reducing family size, and educating women. Kerala is a good example of a place where, in the words of Mills (2003:13) modernity has been used in multiple ways as a 'prescriptive ideal in which the pursuit of modern social forms and institutions is hailed as a movement toward socially valorised goals of "progress", "growth," and "advancement"'. Mankekar (1999:48) analyses changes in family after 1991 and finds that the macro political message was that 'The family ... would acquire a modern lifestyle, and the nation, through the boost consumers would give to the economy, would also "develop" and thus become modern.' At the grass roots level in Trivandrum's middle class neighbourhoods, people associate new forms for consumption with 'being modern'. However, in many consumption domains, for example television and mobility, consumption is changing so rapidly that new ways of consuming quickly get normalised. Chapters 8 and 9 will take up the ways in which consumption is implicated not only in people's efforts to get ahead, but also in efforts to keep pace with changing ideas about what is normal.

Traditional is another of the oppositions to modern, but one with a largely positive connotation in Kerala. The modern, in the form of social reforms and new consumption should displace the backward but should not wipe over tradition. In this book I will show how interest in *both* the modern and the traditional is important to understanding how people from Kerala consume.

## The middle class

This study has been centred in middle class neighbourhoods in Kerala's capital city of Trivandrum. However, I have not been overly concerned with a strict definition of middle class. In studies of consumption, 'middle class' is defined in many different ways, variously using criteria for income, type of job, as well as by the kinds and amount of possessions. For example, Consumer International targets families who can afford basic subsistence, defining an Indian middle class family as one that can afford three meals a day (CI-ROAP 1998:13). *The Economist* (1994 online, cited in CI-ROAP 1998:83) sets their threshold for 'middle class' somewhat higher, defining the middle class as those who can 'afford durable goods such as fans, refrigerators and motorcycles'. These and other definitions based on material criteria amount to a kind of tautology for those who set out to do research on consumption in the middle class: families are defined as middle class because they consume in certain ways; then they are selected for consumption studies because they are middle class. How does one get around this problem and site a study of middle class consumption? O'Dougherty (2002), facing a similar problem in locating her study of middle classness in

Sao Paulo, Brazil, decided to use a mix of local scholars and other informants, supplemented by her own impressions gathered in an exploration of the city and its environs. This is essentially the strategy I employed in Kerala. In locating the middle class, I gave most attention to what the local people designate as middle class neighbourhoods. This designation is based on an assessment of certain traits, including perceived levels of income, education, and capital holdings, mainly in the form of house and land.

A number of recent studies have aimed at studying middle class culture and identity in developing countries (Liechty 2003; Pinches 1999; Walby 1992; MacCannell and MacCannell 1993). Consumption is invariably posed as an emerging element in middle class identity. For example, Liechty (2003), who writes about the middle class in Kathmandu, Nepal, contends that consumption is deeply related to the project of establishing and maintaining middle class identity. Middle class families and individuals work out their identities, and work to maintain them through cultural projects like education and consumption. By contrast, my assessment is that in Trivandrum middle class neighbourhoods, families are not overly concerned with 'debat[ing] the terms of group membership' as Liechty found in Kathmandu. In discussions about their identity and status, and that of others, Malayalee are more likely to use as references caste, job, educational level, extended family, marriage, dowry, or consumption. It is not their fellow members of the middle class, but rather the residents of their neighbourhood, its households and their extended families which are the most important social referents for consumption. The point I want to emphasise is that this book is not about 'how emerging middle classes constitute themselves as cultural entities (Liechty 2003:15)'; rather, it is about consumption among families in middle class neighbourhoods, consisting of families with sufficient economic income and capital to be able to consume beyond their subsistence needs.

## The study

Relying on local scholars, shop owners and acquaintances, the Trivandrum neighbourhoods of Kumarapuram and the adjacent district of Kannamoola were chosen as the primary site of the ethnography. Kumarapuram covers a hilly area on the Western boundary of the Trivandrum city limits. The house where my family and I took up residence was near the top of a hill called Chettikunnu. Our immediate neighbours were mainly Hindu Nair, Hindu Ezhava, and Syrian Christian families.

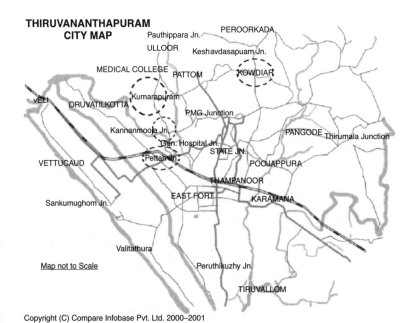

Map 2  Trivandrum (with neighbourhoods Kumarapuram Kannamoola, Pettah and Kawdiar circled).

The most important subjects of this ethnography have been the everyday lives of neighbours and acquaintances. Their experiences form the core of the ethnography. My wife, children and I participated in community life, made friends and developed a social network. We participated with other families in everything from the mundane to the special (festivals, marriage engagements, weddings). I was out and about on my bicycle, or on foot, as much as possible, which made possible spontaneous contact and conversation. Since Trivandrum is a large city, with a population of 800,000 inhabitants, the lives of the families living in Kumarapuram ranged over a large physical space as people went about their commuting, shopping, eating out, visiting neighbours and so on. As ethnographer, I followed these patterns and studied the multifaceted spaces of everyday life.

Over the course of the study, I developed close relationships with several families. I was invited into their lives and participated with them in both mundane and ritual practices. In addition, in depth interviews were done with 80 families in Kumarapuram and other middle class neighbourhoods. I returned to do follow up interviews with many of these families. A few months into our first stay, I began a set of separate interviews with 16 younger people, ranging in age from their mid-teens to mid-20s. For the young women, being interviewed alone would have been unthinkable. I usually met with them in groups of two or three. Finally, a number of families who had recently purchased a household appliance were interviewed. I discussed all aspects of their lives with them, but went into depth on their purchase and the ways the new appliance had affected their everyday practices.

In an effort to get an overview of routines in the home regarding food, cleaning (of self and house), shopping and entertainment, I asked 20 families who had participated in interviews to keep daily diaries. All of the adult members of the household in these families recorded their practices over a period of several days (weekday and weekend) on pre-prepared forms. Altogether 51 adults and a few children – who usually insisted on being included – participated. One of the points of this exercise was to get an idea of who in the household participated in household chores, how often and how much. More importantly, the diaries created a basis for discussion of household routines. In the end, the recorded information in the diaries had only limited value beyond providing a rough idea of how much time people were spending on various activities. However, as I had anticipated, based on previous research (Wilhite and Wilk 1987), the diary exercise was very useful in bringing me into closer contact with families, and it provided a basis for exploring their consumption practices with them.

In the final months of the first year of fieldwork, I conducted a survey, conducted in the form of a set of structured interviews, with 408 households in Kumarapuram, Kannamoola, Pettah and Kawdiar. The questionnaire consisted of 38 questions with fixed-response alternatives. The questions focused on the following: demographic information on the family, factual information on the home and its contents, the division of roles in household chores, and the kinds of products people were or were not consuming. I designed the questionnaire with an awareness of the limitations of this kind of research instrument, especially the tendency of respondents to conceal perceived faults and to exaggerate perceived qualities. In the main, the point was to use the questionnaire to test, and where possible, to quantify insights from the contextually-grounded aspects of the ethnography. The results have been useful in comparing and contrasting household consumption according to characteristics such as religion, caste, age, type of household, and whether a close family member is working outside India.

Another important focus of the study has been the providers and marketers of goods, including retail salesmen, building contractors, architects, and beauty salon operators, as well as subjects seemingly further distant from the everyday lives of people in Trivandrum, multilateral development agencies and NGOs. Print media and print advertising are especially important in this highly literate Indian state, as well as television programming and television advertising. Purchases of televisions and television viewing have both increased steadily over the past decades and watching television has become the most important leisure and entertainment activity for middle class families. As I will elaborate in Chapter 9, many of our neighbours pointed to increased television viewing as constituting the biggest change in social life in Trivandrum of the past few decades. For middle-aged and older generations, television viewing has virtually replaced movie going, and has cut into time spent on other things as well, such as excursions to the beach, strolls around town, and visits to neighbours and so on.

Another home media that I have not given much attention to is the internet. The internet was opened to widespread access in Kerala in 1995. The number of internet cafes increased rapidly in the ensuing decade. There is now a flourishing business in evening internet training schools for adults. Many new businesses in Kerala base themselves on internet outsourcing from Europe, Australia and North America. Access to internet at home has also increased rapidly. In the 408 households that participated in the survey, close to a quarter owned a personal computer and of those, about half had an internet connection; however, I did not dedicate enough time to internet use to be able to make any well-grounded judgments about its effects on consumption. This is a subject that undoubtedly will be important to future research on consumption in India (see Miller and Slater 2000).

## Organisation of the book

Chapter 2 gives a synopsis of important historical background, including the important subjects of family, household organisation and caste. Chapters 3–6 explore the ways that gender roles, marriage and dowry are implicated in consumption. Chapter 7 focuses on home consumption practices and why they are changing; cleaning (body and house), cooling the home and automobility are examined. Chapters 8 and 9 turn to the role of ideas and discourses, examining the ways that political and religious ideologies, as well as development discourses affect consumption. The importance to consumption of the Gandhian legacy of *swadeshi* (indigenous development) is explored, as well as Kerala socialism and Malayalee religiosity. Chapter 9 takes up television, an important conduit of ideas of both local and foreign origin. Attention is given to the popular 'serials', or soap operas, and the advertisements that accompany them. Chapter 10 draws together insights on why consumption is changing in Kerala and on what these insights mean for the theorising of consumption.

# 2
# Global Interchange and Modernising Reforms

This chapter introduces and historically contextualises three important oppositions crucial to understanding consumption in Kerala: global and local, openness and insulation, traditional and modern. Most analyses of changing consumption in India begin with the dramatic effects of the opening of India in the 1990s to transnational corporations, capital and media. The metaphor 'opening' falsely gives the impression that India was closed to global interaction prior to the 1990s. This chapter will explore the importance of several strands of ideas which have moved into Kerala through global contact, their manifestation in socio-political reforms and their importance in changing consumption. These include ideas with their sources in the colonial experience, in Marxism, in international development programmes, and in work migration and media. The openness to these ideas has been juxtaposed in Kerala with a pride in local culture (including language, music, literature and dress) as well as locally entrenched practices concerning gender roles and family.

Ideas from abroad have been one important source of zeal for reform movements of various kinds which have been directed at caste, women and education, each of which has had consequences for consumption. Kerala Historian J. Devika (2002a and 2002b) writes that during and after the colonial era, Kerala governments, social reformists and religious organisations all used terms like 'modern', 'progress' and 'develop' to promote or justify reforms in religion (caste relations, marriage practices, death rituals),[1] education, family (from matrilineal to patrilineal; from joint to nuclear; from big to small), and gender relations (from non-cohabiting to cohabiting conjugal partners). The intention in this chapter is to take up some of the important ways that global exchange and socio-political reforms have affected socio-cultural practices and consumption in Kerala.

Malayalee who wanted to introduce me to Kerala often referred to it as 'God's own country'. This is intended to communicate Malayalee pride in Kerala's cultural and natural heritage, which includes Kerala's unique nature as well as its literature, art, religions and material culture. However, 'God's country' is also referenced to Kerala's social achievements in literacy, education, and openness to the world at large. Whether in their openness to ideas and products from the West (and the developed countries of Asia), or more instrumentally in their pursuit of modernity, there is no doubt that Malayalee have long been open to the World and to change. Malayalee interest in new trends is reflected in the ways that the State of Kerala's Department of Public Relations describes Kerala 'Customs and Manners' on its web site. After a listing of the ways that Malayalee are 'distinct', the following statement appears: 'The outlook on life has changed considerably and new ideas resulting from liberal education have permeated the society.'[2]

References to the changing 'outlook', 'new ideas' and 'liberal (modern) education' are all evidence of the Malayalee concern with being open and modern. On the other hand, Malayalee are also proud of their cultural traditions, including their literature, art, religions and material culture. As will be explored in the coming two chapters, women are seen as repositories of cultural traditions; in their dress, comportment and role as nurturer of family, women should embody tradition. However, the role of women has been a subject of modernising reforms, especially in the domain of education. Kerala women are the most literate and most highly educated in India. The rapid advances in women's educational achievements have drawn attention to Kerala as a model for gender reforms (Alexander 2004; Jeffrey 1992). Kerala women have a higher literacy rate, have achieved higher educational degrees, have lower rates of female infanticide, and have longer life expectancies than women in other Indian states. On the other hand, Kerala's capital city Trivandrum has the highest incidence of domestic violence of any city in India. Suicide rates for men, women and children in Kerala are all among the highest in the world; indeed, the suicide rate for women *is* the highest in the world (Soman and Kuty 2004). In a Ministry of Home Affairs study of domestic crimes, Kerala women reported 'cruelty by husbands and relatives' at a rate that was twice the national average (Eapen and Kodoth 2004:24). These seemingly contradictory data are evidence for multiple pulls on women and on gender roles. The ways that consumption is implicated in these multiple demands will be the subjects of Chapters 3 and 4.

Another subject that is important to understanding almost anything about Kerala – including consumption – is caste. After a brief recounting of Kerala's important historical contacts with global political and economic agents, Kerala's religions and castes are discussed, with emphasis on the ways in which caste relations have changed over the last century and the ways that consumption is implicated in this change. Legal assaults and affirmative action, both couched in the language of modernity, are partly responsible for the change, as are land reforms and a reduction of the economic gap between the two most populous Hindu castes of southern Kerala, the Ezhava and the Nair. This change in caste hierarchy has implications for the ways that consumption is used in social performance, a point which is developed in the final section of this chapter.

## A history of global contact and migration

A favourite joke that Malayalee tell on themselves goes as follows: Neil Armstrong takes his first step on the moon and begins his speech about the 'one large step for mankind'. He stops in mid-sentence as his eye catches a Malayalee tea stand and a Malayalee offering him a glass of *chai*. On the arrival of the next Apollo Mission, the astronauts note that the shop has been taken over by a Guajarati[3] who reports that the Malayalee has moved further out into the solar system. The message this is intended to convey that the Malayalee lifeworld has no spatial bounderies.

Looking at Kerala's history, there is a long tradition of global contacts, long distance economic exchange and outward movement. Kerala's location at the junction of the Indian Ocean and the Arabian Sea has long made it a natural stopping point on the busy trade routes originating in the Middle East and the east coast of Africa from far back into history. There is artifactual evidence that Kerala was visited by Jews and Arabs plying those routes as early as 800 BC. According to legend, Jesus' disciple Thomas followed the trade route when he visited Kerala and made converts to Christianity around 50 AD. Menon (1998) points to evidence that Egyptians and Phoenicians had contact with Southwest India (which today constitutes Kerala) from the 5$^{th}$ century BC. The written record shows that Romans traded with the region from at least the 1$^{st}$ century AD. There is evidence of contact with Chinese traders from as early as the 6$^{th}$ century AD (Menon 1998).

European contacts with Southwest India began with the Portuguese explorer Vasco da Gama, who landed at Calicut in 1498. Portugal rapidly established trade in spices (pepper, cardamom, cinnamon and ginger) and cash crops, including cashew, tobacco, pineapple and papaya. In the

17th century, the Dutch arrived in Southwest India. They added coconut and rice to the established export crops named above and increased the trade in coconut oil. They maintained a dominant position in trade and European political influence until the middle of the 18th century, when they were forced out of the region after a series of skirmishes with Marthanda Varma, the Raja of Travancore (one of the three kingdoms that would be merged into the state of Kerala in the 1950s, along with Malabar and Cochin).

The British began trading along the coast of Kerala in the mid-17th century. Towards the end of the century, the British consolidated its power in Malabar. Cochin and Travancore were able to maintain their sovereignty; however, both kingdoms were forced to sign a treaty with the British in which they were designated as 'subsidiary allies'. One of the conditions of the treaty was that both Kingdoms pay extensive royalties to the British. This was a source of resentment and one of the reasons behind a series of uprisings against British influence in the 1800s. The struggle against the British colonial authority would go on for the better part of two centuries in British Malabar, with support from the Kingdoms of Travancore and Cochin. Resistance took the form of armed uprisings, such as the Mappila uprisings in Malabar,[4] and various forms for political protest from the Indian National Congress (later the Congress Party). Foreign goods and Western materialism were used as symbols of foreign repression in these protests. At a convention of the Congress party in Cochin in 1921, the convention delegates organised a boycott of foreign products. The colonial police intervened and roughly treated the convention delegates. The Indian print media gave significant coverage to the protests, the maltreatment of Indian leaders, and their attack on foreign goods. A decade later, in 1932, Congress organised the 'Second Civil Disobedience Movement', drawing independent activists from around India. Once again, the convention of delegates organised a boycott of courts of law, educational institutions, foreign goods and toddy (palm wine) shops. Partly as a result of the boycott, the British declared the Congress Party to be unlawful and arrested all 400 delegates who attended the convention. These protests strengthened the association of colonialism, repression and foreign goods. The resentment of Western goods and Western materialism would figure strongly in the politics of independent India (discussed in Chapter 8).

Kerala did not become an Indian state until a decade after India's independence in 1947. After independence, the first step toward Kerala statehood was the merging of the Kingdoms of Travancore and Cochin in 1949. The Maharaja of Travancore became the Raja of the new kingdom called Travancore-Cochin. The Legislatures and Ministries were combined. The seat of government of Travancore-Cochin was placed at Trivandrum and the High Court located at Ernakulam, just outside Kochi. In 1956, the Raja of Travancore abdicated his thrown, and Malabar merged with Travancore-Cochin to form the state of Kerala.

## Kerala socialism

Zealous social reform, anti-colonial sentiment and educational advancements all contributed to the growing popularity of Marxism in Kerala. According to Jeffrey (1992), by the mid-1940s left-leaning student organisations, largely made up of Nair men became a potent force for change. They organised frequent strikes and protests and were behind attacks on political leaders. Many of the men, who had earlier supported Kerala religious reformers such as Shree Narayana Guru or Vivekanada, gave their allegiance to the Marxist movement and helped to form the Kerala branch of the Communist Party of India (CPI).[5] One could say that after the British colonial influences, Marxism ranks as one of the most important outside influences on the political and economic development of Kerala.

In the first general elections in Kerala in 1957, a coalition of left political parties was elected and formed the first government. The government's Chief Minister was a member of the Communist Party of India (CPI), making Kerala's government the world's first elected government to be lead by a Marxist. According to Franke (2002), the path to Marxist governance was strongly influenced by the communist revolution in the Soviet Union. The Indian independence movement was transfixed by the struggle of the Russian oppressed and their dramatic assumption of power. In Kerala, Marxism appealed to 'an ideological void keenly felt by thousands of literate, alienated people (Jeffrey 1978:78)'.

After the initial elections and the success of the CPI in gaining control of the State government, land reform, a typical socialist project, was one of the most important priorities of the first Kerala government. The first land reform legislation was approved by the Kerala legislature in 1959, but its implementation was blocked by the Indian government. Kerala's legislature then modified the reform in 1960, and it was finally approved by the Indian government; however, its implementation was blocked once again because the Left political coalition was voted out of power in Kerala for two brief periods in the 1960s. The Kerala Land Reform Act was finally implemented in 1969, two years after the CPM-lead United Front recaptured leadership of the state government (Raj and Tharakan 1981).

As a formal part of the implementation of the Act, the Kerala government appropriated all land holdings and immediately redistributed them. After the reforms, the maximum permissible size of a holding was 20 acres. Tenants on landed estates were awarded the land that they had previously leased. Lower caste agricultural labourers were granted deeds to their houses and to 1/10$^{th}$ of an acre around them. This land reform significantly redistributed capital and income, and contributed to the growth of Kerala's middle class. Subsequent redistributive policies and other social reforms promoted by Kerala governments have been dubbed the Kerala Model of Development. The Kerala Model and its relationship to consumption will be developed in Chapter 8.

It is important to keep in mind that because Kerala is a state in India, it is subject to changes in Indian national policy. Since 1991, Kerala socialism has been literally enveloped in a national policy that 'opened' India to global capitalism. An important point of the discussion thus far has been to demonstrate that the metaphor 'opening' is not really appropriate when applied to Kerala. Kerala had long been open to foreign trade, political influence and ideas from the outside world prior to the 1990s. However, the post-1991 changes represent a different order of opening in which the policies, products, and the ideology of transnational capital, held at bay for four decades after Indian independence, were allowed in. The implications of this 'opening' for consumption are explored in Chapter 8.

## Reforms in family and gender relations

Another target of modernising reforms has been kinship, household and gender relations. In the 19$^{th}$ and early 20$^{th}$ century a series of reforms took aim at the matrilineally organised extended family. Kerala's *marumakkathayam,* involving matrifocal residence and matrilineal inheritance was practiced by about 60 per cent of the population in the early 19th century (Gough 1962a).[6] Nair families lived in joint family households consisting of matrilineal kin (called *taravad*). A form for marriage was practiced (called *sambandham*), but husbands did not cohabitate with their wives. Men lived in the *taravad* of their own matrilineal related kin. The senior male in the *taravad*, designated the *karnavar*[7] controlled household affairs. Household consumption was regulated by the *karnavar* and other senior members of the joint household, a practice that was to continue well into the 20$^{th}$ century. The *karnavar* allocated food (rice, vegetables, coconut oil, milk, and other stores) and clothing to the members of the household. He also selected the marriage partners for female members of the *taravad* and exercised the role normally assigned to the father in patrilineal families. As Gough put it, 'He exercised the rights and obligations over children that normally accrue to the legal father in the higher patrilineal castes of India (1962a:352).'

*Marumakkathayam* was the target of a sustained assault by social and religious reform movements of the 19th and early 20th centuries. Religious reformers with their inspiration in both Europe and other parts of India characterised *sambandham* as immoral. The reformers quoted practices in the 'civilised' world (Europe) and insisted that it was a man's right to bequeath his self-acquired wealth to his wife and children instead of to his *taravad*. They championed the moral superiority of the patrilineal system. Because *sambandham* was not conducted in either a church ceremony or a civil ceremony, and could be dissolved by mutual consent of partners, its practitioners were characterised as 'promiscuous' and the children of *sambandham* relationships were characterised as 'bastards'. Saradamoni (1999) writes that European-educated Kerala men were some of the most ardent reformers of *marumakkathayam*, especially junior male members of *taravad* who had much to gain from an acknowledgement of their rights as husbands (among these would be control over their incomes and inheritance). These men were some of the most tenacious opponents of the *karnavar* after the mid-19th century.

The reformist thrust contributed to a series of laws in the early 20th century which changed rules about family, property and inheritance. In Cochin, a law was passed in 1920 making polygamy, which *sambandham* was considered to be an example of, illegal (Puthenkalam 1977). In Trivandrum, the Nair Act of 1912 declared that half of the self-acquired property of the male should no longer go exclusively to his sister's children, but should instead be equally shared with his own children. The second Nair Act of 1925 went further. It deprived nephews of all claims to the property of their uncles and provided for partitioning of joint *taravad* holdings. The Ezhava (1925) Act and the Nanjanad Vellala Act (1926) provided similar changes in the inheritance laws (Menon 1998:408). These acts legally recognised the conjugal family and set out relations of protection and dependence between husband and wife and between father and children. Guardianship of the wife and children was legally ceded to the husband. Divorce, which was settled informally under *sambandham*, was made a subject for the courts.

There is a record of bitter and lengthy litigation over the break up of the *taravad* property.[8] Satira, one of our elderly Nair neighbours who experience life in a *taravad* tells how her uncle, the last family *karnavar*, was forced to break up their *taravad* property and in her words to 'destroy everything'. According to her, this left her mother and uncles sad and disoriented. The younger brothers were forced to move to other parts of Kerala. The social reformer Mannath Padmanabhan describes extended families of the early 20th century as 'spheres of war' (sited in Jeffrey 1992: 34). He exhorted what he refered to as 'alienated' men to find jobs, earn their own income, and take responsibility for their wives and children. This call for reform became an issue for Marxist and socialist political activists such as E. M. S. Nambutiripad, who exhorted men to 'rise up and assume the role of producer and provider for their families (Devika 2002b)'.

These significant changes in family over the course of a few generations have relevance for consumption today. The most important perhaps is the restructuring of household from joint family households to nuclear households, increasing the number of houses, household appliances, furnishings and so on for a given extended family and affecting the organisation of housework. The strongly gendered division of roles in both marriage-partner selection and in housework in the matrilineal *taravad* were carried over into nuclear, patrilineal households. The ways this has affected consumption will be developed in Chapters 3 and 4.

## Caste reforms and the blurring of caste hierarchy

In Kerala, religion and caste are important in everyday practices and in many of the important cultural and social aspects of Malayalee lives, including marriage, political affiliation, and religious participation. In an earlier time, caste was determinant in virtually every aspect of life, including type of work, access to education and socio-economic class. The point of this section is to argue that when it comes to employment, education levels, economic status, where people live and the ways they consume, the caste reforms of the 20th century have lead to a blurring of the relationship between caste-rank and socio-economic class. This socio-economic levelling has affected consumption in important ways.

Some brief historical points on Kerala's religions and caste system are necessary prerequisites for understanding the role of caste today. First, Hindus constitute only about 60 per cent of the population of southern Kerala. Many forms of Christianity are practiced in Kerala; altogether about 30 per cent of the population is Christian. Muslims constitute about 10 per cent of the population. Christian denominations have very different histories and have achieved different rankings in the Kerala social hierarchy. Syrian Christians trace their roots back to the Christians from Armenia and Syria who settled in Kerala in the 3rd and 4th centuries AD. They established themselves and traders, merchants and landholders and were equated a place in the social hierarchy approximate to that of the upper caste Hindu Nair. From the 15th century, European Christian missionaries brought their version of Christianity, Roman Catholicism, to Kerala, but their attempts at converting Syrian Christians met with limited success (Dempsey 2001). Most of those who today are referred to as Latin Christians are descendants of fishermen who were converted by Catholic missionaries. In the last two centuries, protestant missionaries have also converted many of the tribal peoples and members of the scheduled castes to their versions of Christianity, resulting in a broad tapestry of Christian denominations in Kerala (see Kariyil 2000 for an overview).

As a result of the extensive trade around the Gulf, many Arabs had settled in Kerala prior to the arrival of Islam in the 7th century. After the 7th century, a number of Hindus in northern Kerala converted to Islam. Muslims took positions in the Malabar navy and many became merchants and traders. In the northern part of what is today Kerala, they had a social ranking similar to that of the Syrian Christians (Gough 1962a:313). In the 18th and 19th centuries, northern Kerala was invaded by the Muslim Hyder Ali. Many fishermen and tribal peoples converted to Islam, creating a large population of Muslims whose economic and social circumstances were similar to that of the lower Hindu castes (Menon 1998).

For Hindus, the ideology of caste ranking is established in the Vedic doctrines.[9] Members of the Brahmin caste are said to have achieved the highest spiritual stages of human form. In Kerala, from at least the 8th century, there was a strict caste hierarchy with the Nambuthiri Brahmin caste at the top (Iyer 1981; Gough 1962a).[10] The Nambuthiri Brahmins assumed roles as religious leaders, political advisors and intellectuals. Many of them accumulated large land holdings. Gough (1962a) likens the role of the Nambuthiri in early 20th century Kerala to that of the priests of the medieval Catholic Church in Europe. However, the power of the Brahmin began to decline in the later 20th century. One reason for this decline was the strong social and religious reform movements in Kerala in the 19th and early 20th centuries. Caste privilege was one of their targets. As the caste privileges of the Nambuthiri were undermined and their land holdings broken up through land reforms, members of the Brahmin caste were forced to take jobs which formerly had been done by other (lower) castes.

The Nair caste is the second ranked caste in the traditional hierarchy. In the 18th and 19th centuries, members of the Nair caste held positions in the military and in government. Many Nair families had extensive land holdings. Most were well educated. In the early 20th century, university student bodies were dominated by men of the Nair caste. However, land and legal reforms of the late 19th and early 20th centuries directed at marriage and inheritance weakened the economic situations of many Nair families. Some were forced to move from their estates and to find employment. Many took jobs in higher education and in the public schools. The economic status of the Nair began to decline. Gough (1962a: 35) wrote that 'many commoner Nayars in both Central and North Kerala were obliged to sell their small holdings to larger capitalists. They became salaried workers in agriculture, bailiffs, school teachers, cooks, clerks and government messengers and chauffeurs.'

Gough claims that by the late 19th century, Nair caste members began to do work normally done by members of lower castes. Fuller (1991) and Gough (1962a) both identified what Gough called 'Nayar menials' who were domestic servants for members of their own and other castes. Members of these lower Nair sub-castes did not mingle with or intermarry with the upper Nair (Gough 1962a and Puthenkalam 1977). Thus the socio-economic status of the Nair caste has long been splintered and distributed across Kerala's socio-economic spectrum. In Trivandrum today, many people still make a distinction between 'upper' and 'lower' Nair. It is not uncommon for Nair women to work alongside women from the Ezhava caste and Latin Christians as domestic servants in middle class and wealthy households.

The Hindu Ezhava caste is officially categorised by the state as OBC (other backward castes). Until the mid-20th century, the Ezhava worked mainly as agricultural workers and producers of palm wine (*toddy*). Below them in the hierarchy were the *Pulaya* and *Pariah* who were considered to be outside (or below) caste classification and were regarded as polluting to members of upper castes. Neither the Ezhava nor the other lower castes were allowed in temples or to touch the person, food or serving instruments of a Nambuthiri Brahmin. Even the shadow of a person of low caste was considered to be polluting and was to be avoided by Brahmins. As Brahmins moved along the roads, their progress was heralded by an attendant in order to signal to lower caste or non-caste members that they should move out of the way.

During the period in which the social and economic status of the Nair was declining, the status of the Ezhava was rising. The lower-caste status of the Ezhava and the OBC were the targets of caste-based reform movements from the late 19th century. Sri Narayana Guru's reform movement, begun in 1903, aimed at giving members of the Ezhava caste the right to enter temples. This was realised in 1936, when the Maharajah of Travancore proclaimed temples open to peoples of all castes. The Ezhava were able to acquire land and accrue political leverage due to legal and political reforms favouring lower castes.

Osella and Osella (2000:52) cite an example of how caste and social status were shaken up in the early 20th century. An Ezhava man, Koccu Kunja Channar, bolstered by the commercial successes of his father, built up considerable wealth and land holdings in Travancore. He had acquired 12 elephants, an important indicator of status. Each elephant had its own *mahout* (caretaker). One day Channar rode his elephant to the village temple, assisted by his *mahout*. Channar, because he was a member of a lower caste, was forced to dismount and walk through back alleys to the temple; his salaried Nair employee was permitted by his upper caste status to ride down the main thoroughfare to the temple on Channar's elephant.

Osella and Osella devote their treatise on social mobility in Kerala to the story of the upward social mobility of the Ezhava caste. After the creation of Kerala, access to primary and higher education, including universities, increased dramatically for the lower castes. Allotment programmes gave members of Ezhava and other 'scheduled' castes quotas in universities and in selected government jobs previously held mainly by members of the Nair caste. With their ethnographic focus on a rural village, Osella and Osella show how access to education, land and government jobs has contributed to raising the socio-economic fortunes of the Ezhava. They argue however, that the Ezhava are still regarded as having lower social status than the Nair, and that consumption is one strategy used by Ezhava families to progress socially.

The results of my research show quite clearly that social and economic reforms have had the consequence of levelling differences between Ezhava and Nair castes in Trivandrum, at least measured in terms of income, education and possessions. This parity is also reflected in the settlement pattern in Trivandrum middle class neighbourhoods. Nair and Ezhava are mixed in each of the four Trivandrum neighbourhoods studied. Figure 2.1 shows the distribution of Ezhava and Nair families in these four middle class neighbourhoods.

This mix of Ezhava and Nair in the neighbourhoods of Kannamoola and Kumarapuram is one indication of a change in the relationship between caste and social status from former times, when upper caste settlements were segregated from those of lower castes. The remnants of this segregation still exist in Kawdiar and Petah. The population of Kawdiar is still dominated by Nair families, while Pettah is populated mainly by Ezhava families. However, the average household incomes the

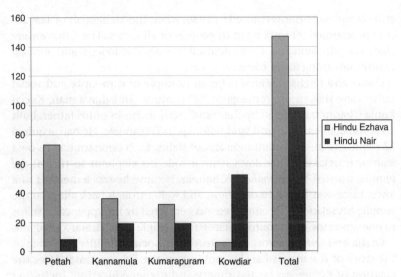

*Figure 2.1* Distribution of Hindu Nair and Ezhava households in four middle class Trivandrum neighbourhoods, based on a survey of 408 households

Nair and Ezhava castes are almost identical – 8200 rupees for Ezhava and 8260 rupees for Nair.[11] In Kawdiar, where Nair dominate, the average income is actually less than that in Kannamoola, where there are more Ezhava families than Nair. Pettah, with the greatest proportion of Ezhava families, has a higher average household income than that of Kumarapuram. These income comparisons show that differences in economic status between families of the two castes are today insignificant.

On a given street, such as the one on which my family resided in Kumarapuram, Ezhava and Nair homes are interspersed along the street. Children of the two castes play together and attend the same schools. Friendships and social interactions span caste difference. This is evident in the student populations at schools and universities, in work places, and in public spaces such as on public transportation and restaurants. Members of the older generation of Nair occasionally made statements that revealed stereotypes they held about Ezhava, such as their purported problems with alcoholism and their illegal sales of alcohol. These stereotypes are not shared by most young people. For the vast majority of young people I interviewed or interacted with, caste was not a part of their vocabulary of social distinction. Rija, a 24-year old Ezhava, told me she had friends of all castes; in fact, she claimed that she only thought about caste when deliberating her choice of a husband (marriage practices and partner selection figure into the discussion in Chapters 3, 4 and 5). 'In my group of friends we are not concerned about the caste. It doesn't matter whether you are from this or that caste.' She lived much of her childhood in Northern India, where her father worked for many years. She said that there she mixed with other Malayalee without ever thinking about caste differences. 'It was only after I came to Kerala that I came to know that Hindus had a caste called Nair in Kerala. When I was there (in the North) there were only Tamalis (from Tamil Nadu), Malayalee and so on.' Our middle-aged Nair neighbour, Meena made the same point and illustrated it with the example that people no longer worry about the caste of the person sitting next to them on the train or dining at the next table in a restaurant.[12]

Both educational and economic reforms have contributed to the levelling of economic differences between castes. The two castes have also reached parity in educational achievements. The following figure compares the educational achievements of Nair and Ezhava female heads of households surveyed.

Figure 2.2 shows only small differences in the education levels of women of the two castes; in fact, Ezhava women have achieved, on average, slightly higher educational degrees than Nair women have.

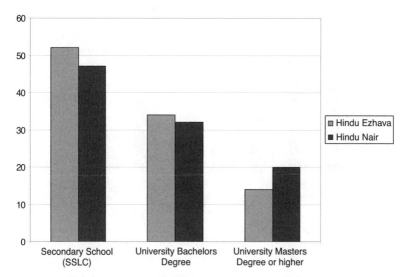

*Figure 2.2* Percentage of female heads of households of Ezhava and Nair castes having completed secondary school, bachelors and masters degrees, based on a survey of 408 households in four Trivandrum neighbourhoods

A final measure of socio-economic parity is ownership of appliances and motor vehicles. Figure 2.3 compares ownership of cars, motor cycles and selected appliances in Ezhava and Nair households.

The figure indicates very small differences between castes in ownership of these items.

The economic and educational parity between these two castes has reduced the power of caste as a determinant in the social hierarchy. Osella and Osella (2000) found that in a rural Kerala village, the Ezhava caste still exercise a form for cast emulation through consumption and that 'progress should be demonstrated and publicly acknowledged by keeping up with levels and styles of consumption of those to whose status and prestige one aspires: the Nayar and Christian middle classes (2000:124)'. A central point resulting from this study in urban Trivandrum is that socio-economic levelling has weakened the caste hierarchy; as a result, caste emulation and caste mobility are no longer important motivations for changing consumption. Liechty found a similar development in Nepal's capital, Kathmandu.

> Today, Kathmandu's squalid squatter settlements include Brahman families of the highest ritual ranks (Gallagher 1991), while members

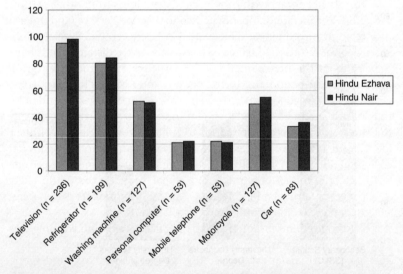

*Figure 2.3* Ownership of household appliances, cars and motorcycles among Hindu Ezhava and Hindu Nair families (percentage of families owning based on a survey of 245 households in four Trivandrum neighbourhoods)

of the once despised and marginalized ethnic groups preside over vast transnational business empires...In the new economic order, social mobility does not occur at the caste level; whole caste communities do not move up or down. Rather, families and (less often) individuals move into positions of greater or lesser socio-political power in which the language and practice of class more adequately conveys shared interests and values (2003:64).

The waning importance of caste mobility and caste emulation in consumption has been noted by Indian market researchers as well. Bullis, the author of *Selling to India's Consumer Market*, wrote (1997:65) that caste is no longer an important 'market segment'. However, caste is still of paramount importance in marriage, where caste plays an important role in marriage and the choice of marriage partner. However, it is important to note that none of the Ezhava families with whom I interacted expressed any interest in marrying 'upwards' into a Hindu Nair family.

## Conclusion

It has been important to establish that Kerala and its peoples have had a long history of contact with people, goods and ideas from other parts of the world. The metaphor 'opening', used to describe Kerala's (and India's) new relationship to global markets in 1991 is deceptive. Kerala's globalised economy and the Malayalee fascination with 'distant' goods have their roots in the pre-colonial era. Nonetheless, it is true that during India's colonial period and in the decades after Indian independence, access to foreign goods was restricted. This restriction was extended by the political architects of independent India, who were concerned with the threats of Western materialism to India's economy and to Indian identity. In 1991, this concern and with it the barriers to transnational capitalism dissolved. Politically, the consumption of foreign goods became an indicator of India's importance as a global player on the World stage. At the level of family and individual, consumption took its place alongside education and job as indicators of social status.

It has also been important to explore Kerala's fascination with reforms and to outline the ways that social, religious and political reforms have affected the social landscape. Family, gender, caste have undergone significant changes over the course of a few generations. A socialist political agenda has been both a result of and catalyst for reforms. Changes in each of these domains – political economy, family, gender and caste – are relevant to understanding changes in consumption. The reflections on caste in this chapter will be important in interpreting the roles that social distinction and performance play in consumption. In coming chapters I will argue that family and neighbour have taken the place of caste as important social referents for consumption.

# 3
# Women in a Bind: The Crucible of Marriage and Dowry

In this and the next two chapters the focus is on family and gender, and the ways consumption is implicated in both. As I engaged with consumption in Kerala, it became increasingly clear to me that gender roles and relations are essential to understanding changing consumption. As Mills (2003:17) writes,

> Gender categories present powerful tools for understanding social life in almost every culture. In any given setting, gender constitutes a complex and historically constructed system of beliefs and social relations...When new social forms and practices arise, which alter or challenge conventional gender relations and beliefs, the cultural conflicts that result may be felt as profound stresses or tensions either of personal identity and day-to-day relations or within the social order itself.

The view from inside the Kerala family reveals that there are social and family pressures at work that are relevant to understanding consumption practices. One of these stems from the double-edged demand that a woman should exemplify both the modern and the traditional. She should be modern in the sense that she should be educated and cognizant of changing ideas about beauty and fashion, but she should also look and submit to the nurturing role regarding home and family. The strategies that women use to negotiate these multiple demands affect consumption in important ways that will be discussed in this and the next chapter.

This chapter will focus on the period in which an unmarried woman is a bridal candidate; a period in which she must demonstrate that she is capable and willing to take on the role of wife. Through her dress and conduct, she must show a willingness to take full responsibility for housework and family (husband, children and husband's lineal relatives), whether or not she takes on work outside the home. I will explore the ways that these pressures affect consumption in this pre-marital phase.

The choice of a marriage partner in Kerala is a family matter. The search for candidates is directed by the senior males (father, uncles and older brothers). The marriageable man or woman may suggest a candidate to the parents, but in most cases this is regarded as no more than a nomination. The families of the bride and groom evaluate the two candidates in much the same way as they would candidates suggested from some other source, such as a member of the extended family or a marriage broker. Ninety-five per cent of the 408 households who participated in the survey reported that the parents actively participated in the choice of marriage partners. This is supported in a study by Murickah (2002) of the marriages of 400 university-educated Kerala couples in the cities of Trivandrum, Cochin, Calicut, Kottayam and Trichur. Murickah found that parents had been active in the choice of partners in 97 per cent of marriages.

Candidates for a bride or groom come almost exclusively from the same caste, or in the case of Christians and Muslims, from the same religion. This contradicts a popular discourse, promoted in popular magazines and newspaper stories, which suggests that cross-caste marriages are becoming more common. Cross-caste marriages make dramatic reading because they involve stories of people who go against the social grain. The reality is that cross-caste marriages in Trivandrum are very rare. In 98 per cent of the 408 households surveyed, the marriages were endogamous (same caste). This finding is consistent with those of other recent surveys. In Lindberg's (2001) study of cashew workers in rural South Kerala, 92 per cent of marriages in 506 families surveyed involved same-caste marriage partners.

Dowry is an integral part of marriage transactions. The family of the bride provides dowry in the form of jewellery, land, money and household goods (including cars) with the bride. Changes in the constituents of dowry over time will be taken up in Chapter 5. The important point to be underlined here is that in its 19$^{th}$ century version, dowry for Nair and some Ezhava families was viewed as property of the bride and intended for her use – called woman's wealth (*stridhanam*) (Gough 1962a; Fuller 1976; Iyer 1981). The things passed along in dowry were intended to be of use for the new bride. According to Satira, an elderly Nair grandmother, the bride's family in the early 20$^{th}$ century 'would only put a gold chain around the girl's neck and give a set of clothes called the *pudava.*' The amounts and constituents of *stridhanam* were regulated by caste and social standing and thus not negotiated (Eapen and Kodoth 2001).

The practice of *stridhanam* began to change in the period of family upheaval during the 20th century. For Nair and other matrilineally organised families, the source of marriage partners was no longer their matrilineal kin. Dowry developed into a negotiation between families with no kinship bonds and in many cases no prior contacts. A young Nair man, reflecting on changes in dowry told me, 'Before it was something that was about confirming social status. But later it just came that if there is no dowry you can't marry the girl. They (the groom's family) started looking into the dowry more than the girl.' The economic dimensions of dowry began to increase significantly by the 1960's (Lindberg 2001).[1] In addition to material goods, men want their wives to be educated. Indian sociologist Srinivas wrote that universities in South India are 'respectable waiting places for girls who wish to be married (1996:153).'[2] His claim is that achievements in higher education for women have actually been bolstered by the prospect of finding husbands and reducing dowry. Dowry and educational demands on brides result in long partner searches and later marriages. Kerala women marry on average at 22.7 years of age, compared to the national average of 17.7 years of age. Nationally, 82 per cent of women aged 20–24 are married; in Kerala, only 56 per cent of women in this age group are married (India Census 1991).

## Social scrutiny

During the period when unmarried women are bridal candidates, ranging from the late teens to late 20s, they are subjected to intense social scrutiny by their family, neighbours and families of potential husbands. Sabena, a 22-year old Ezhava, expressed the stress and frustration that many young women feel:

> Here living life as a girl is quite difficult, because you have to keep in mind that whatever you do there are so many other people watching. You have to be socially correct in whatever you do. The same thing done by a boy does not attract much criticism, but if it is a girl then you have to be careful, you have to watch what you are doing and that really doesn't bring you much freedom...Freedom I don't have.

Sabena, speaking English, used the expression 'socially correct' to express something in Malayalam called *swabhaavam* (Joseph 2002:90). As Sabena implies, *swabhaavam* involves rules concerning appearance but also concerning visibility and movement.

## Mobility

When they are at home, young women are expected to remain hidden from view from the street. They should maintain a discreet distance from the front windows and doors of the house. If a male visitor enters the home, the young women should only come forward if beckoned by a senior family member. When she leaves the immediate neighbourhood to shop or to engage in social or leisure activities, she should be accompanied by a family member. The concern about unchaperoned communication with men is behind restrictions on the use of the telephone. Many young women told me that they are not left alone to make or receive phone calls. The advent of mobile telephone makes supervision difficult. Very few young women are allowed to have one.

As Sabina indicates, the same restrictions do not apply to young men. Young men whose families can afford them allow them to have mobile phones. Young men do not have as many restrictions on their movements as young women. The sons of middle class families unproblematically drive motorcycles and their parent's car. I only encountered one young unmarried woman, 24-year old Savita, who regularly used a motorbike. She recounted many scornful reactions from neighbours, including young boys who threw rocks as she drove by. Savita had lived outside Kerala most of her life and only returned when she was 20. She is one of the few young women I encountered who consciously 'resists' the restrictions of *swabhaavam* by conducting herself in ways that challenge it (Ortner 1995 and 1999). Driving the motorbike is one example; she also frequently dresses in casual clothes, wears make up and moves unaccompanied around town. Savita is well aware that her conduct will affect her prospects for finding a husband, but she told me wryly that she would be happy to find a soldier or sailor who would take her away from Kerala.

Trivandrum may be more conservative than other Kerala cities concerning gender relations, but Joseph (2002) found similar strict restrictions on young women's appearance, movement and 'conduct' in Kerala's commercial capital Kochi. Based on extensive interviews with young women she found that 'there was little question of their being allowed to go anywhere else except in the company of one of their family members, whereas their brothers were free to roam about whenever and wherever they pleased (2002:90)'. Joseph's conclusion was that 'Young women in Kerala are obviously aware that the restrictions they are routinely subjected to are meant primarily to safeguard their apparently all-important reputations. They suggest that as far as society is concerned the most valued attribute in a girl is what is known as "good character".'

Based on a review of recent studies around Asia, gendered restrictions on movements and social space are characteristic in a number of settings. For example, Mills writes that in Thailand, 'Special mobility is perceived as a natural and appropriate characteristic for men, while women's bodies and their movements are subject to far greater restrictions (2003:18).' Liechty writes that in Kathmandu, young middle class women who wear 'excessive ("over") make-up and short skirts' are called *chara*. 'At best, *chara* implies "free-roving," but its main connotation is those who function outside of social controls or norms (2003:72).'

## Dress

The exercise of good character is directly implicated in the consumption of mobility and telecommunications, but more importantly in body, dress and beauty. Hansen studied clothing consumption in Zambia and wrote that 'the dressed body both conceals and reveals deeply entangled issues about gender, sexuality, and power' (2000:207). In Trivandrum, these entangled issues have applied to both genders in the past, but today, much of the attention concerning body and dress is on women. Redu (Hindu Ezhava) and Gurt (Syrian Christian) discussed with me young men's expectations concerning young women's appearance and dress.

> Gurt: Most of the girls here (in Kawdiar) will not wear casual clothes; it is not good, because most of the people here regard casual clothes as bad.
> Redu: Yeah, bad!
> Gurt: It's not good, so most of the girls don't do that. Most people would think she is 'different'.
> Redu: Or she is trying to show off.
> Gurt: People will ask 'Why is she dressing like that?' Dressing in casual clothes is really not good. People might think they don't cooperate.

By 'casual clothes' Gurt means Western-style clothes that reveal too much of the body's shape or skin. By 'they don't cooperate', he means that by wearing casual clothes young women push the limits of *swabhaavam*, and by doing so they cast doubt on their willingness to fulfil their future role as dutiful wives. Few women of marriageable age risk wearing Western-style clothing, except in certain safe zones such as university campuses and a few cafes and restaurants. In Trivandrum, the most popular such cafe is called *Ambrosia*; here, high-heeled shoes, t-shirts, miniskirts and jeans are common. Young men and women share the same tables and flirting is common. However, on leaving young people of both sexes separate into same sex groups.

It is common for young women to change into more conservative clothes before they leave the cafe. Many young women told me that wearing 'casual clothes' back into their own neighbourhoods would only bring them trouble. Seventeen-year old Preethi told me, 'I used to wear anything and everything. But there was some trouble on the road; people call out things and neighbours say some bad things.' Ped, a 20-year old Ezhava said, 'I couldn't be comfortable wearing jeans if I had to drive back after 7 PM because people have this habit of looking at you and thinking "My God, something odd, something weird".'

Nineteen-year old Vibek expressed something similar to Gurt on how he interpreted 'casual clothes', or in his terms, 'modern clothes':

> They can wear modern clothes, but depending on the circumstances and where you are going. There are some places I prefer girls to stick onto traditional clothes.[3] I think it is the case that they (men) respect girls for what they are when they are friends, but when they come to their own choice about marrying a girl, they will always prefer a girl who is not so fashionable (his fiancée, Preethi, interjects 'Yeah!') – who is more conservative and traditional and they are sticking on to that.

Preethi followed this up by saying, 'They (the boys) like girls who are a bit broad minded and all but when it comes to their own girls they are very, very possessive and they want her to be traditional.' Ped, a Hindu Ezhava woman of 20, who had spent much of her childhood in Saudi Arabia (her parents were work migrants who retired in Trivandrum.) said, 'People have the general idea that girls who wear a little bit of casual clothes, guys (potential husbands) won't come.'

It was quite clear that young teenagers who have not yet reached the mate-seeking age have much more latitude in experimenting with clothing and make-up than older teens. Vibek and Preethi met when he was 18 and she 16 at the 'Alliance Française', where they took French-language courses. Initially, she occasionally wore jeans or a Western blouse and skirt to classes. As they began to get interested in each other,

her clothing style changed. By the time she was 18 she wore exclusively Indian style clothing (either a *churidar-kurta* or *sari*).

The *sari* and *churidar* are the clothing styles most men refer to when they talk about traditional women's clothing. The Kerala *sari (mundum neriyathum)* was the public dress for women of all ages and castes until the 1980s. It consists of two cloth pieces that cover the upper and lower half of the body. While it may be arranged to reveal the stomach and back, the *sari* covers the torso and hips in layers of cloth which obscure the shape of those parts of the body. In the late 1980s the *churidar* (pants) and *kurta* (blouse) took its place alongside the *sari* as an acceptable Indian dress style for young women. The *churidar-kurta* is often worn with a shawl-like garment. The shawl can be positioned in different ways to cover or reveal the head, neck and arms in accordance with the social circumstances. Recent versions of the *churidar-kurta* are more tightly fitting. The kurta sleeve lengths are getting shorter. A decade ago, sleeves covered the entire arm. *Kurta* sleeves now reach only as far as midway down the bicep. For the daring, there is now a sleeveless version. A canvassing of retail stores revealed that these sleeveless *kurta* come with a kind of social insurance pinned into the inside of the garment, namely sleeves, giving the purchaser the option to retreat to a safer social position by sewing them into place. There is no doubt that the more revealing designs reflect a gradual change in ideas about the ideal female body and the way it ought to be displayed. Nonetheless, even though recent styles are daring, they still signal that the bearer is cognizant of the importance of Indian tradition.

The modern-tradition conundrum is much less pronounced for men. Where young men are concerned, Western-style clothing is not associated with 'bad character' as it is for women. Most young men in Trivandrum today regularly and unselfconsciously wear Western-style shirts and trousers outside the home (see Plates 2 and 4). Tarlo (1996) traces the use of Western clothing by men back to the colonial era, when elite and educated men used it to distinguish themselves. In a reversal influenced by Gandhi (discussed in Chapter 8), Western clothing for middle class and elite men went out of favour in the early 20$^{th}$ century. By wearing the traditional *lungi* and *dhoti*, the bearer made a cultural and political statement about his interest in supporting basic Indian cultural values. In recent decades these political motives and the cultural/moral implications of dress styles have all but disappeared. Many middle class men in Trivandrum, mainly middle-aged and older, still wear the *lungi* in the home and around the neighbourhood, but mainly for its comfort (see Plate 5). The *munde,* a more formal version of the *lungi,* is used sometimes on formal occasions by middle-aged and older men. Ped (Ezhava woman in her mid-20s) remarked that the *munde* is rarely worn by young men today. She stated emphatically that 'Most of the boys of our generation do not even know how to wear a *munde*! They are not even confident to wear it.' Sabena, a young Nair woman, attributed the freedom young men have in their choice of dress to their exemption from social scrutiny and the traditional-modern dilemmas constantly confronting women. She told me 'I don't think there exists a social definition of how a boy should conduct himself. I don't think they are ever questioned about anything they do.'

## Beauty

While female beauty ideals concerning shape and clothing are slowly changing, an important ideal for both men and women that has long historical roots is fair complexion. Osella and Osella (1996 and 2000) cite the widespread belief in rural Kerala that dark skin tone is a characteristic of lower castes. From the early 19th century women have used powders, creams and treatments to lighten their skin. The first mass-produced fairness creams appeared in the early 20th century.[4] Today, fairness is one of India's biggest businesses, and Kerala women are its biggest consumers. Kerala with 2 per cent of India's population consumes 40 per cent of all 'fairness products' consumed in India (Phookan 2004). Two successful Indian cosmetic companies, Lakshmi and Fair & Lovely make huge earnings from their fairness products (Phookan 2004). In the past decade multinational corporations (hereafter TNCs) have also increased their sales of fairness products; examples are Unilever (through its subsidiary Hindustan Lever Limited), Nivea (through its Indian distributor J. L. Morison), and Ponds. Fairness products are amongst the most heavily advertised products in Kerala (Chanda 1991). Many of the advertisements are obviously aimed at young women and blatantly portray the social advantages of being fair. Fairness is also a subject of many of the popular evening serials produced in Kerala (explored in Chapter 9).

Laila is the proprietor of the beauty salon called 'My Fair Lady' (see Plate 1). Her biggest business is in the sale of fairness products and fairness treatments and her biggest clientele are the mothers of young women who are either preparing to meet a potential mate; are preparing for their engagement ceremony; or, are preparing for the wedding itself. Laila told me that the mothers almost always accompanied their daughters.

Laila says that women want a 'natural look', meaning one that enhances, or produces fairness without looking gaudy or artificial. Laila produces

**Plate 1** My Fair Lady Beauty Parlour, Trivandrum

this look with carefully applied layers of creams and powders. She lamented that her 'treatment' is getting competition from chemical treatments the effects of which last for days. She showed me a brochure describing a chemical fairness treatment (the 'golden face' treatment) costing several thousand rupees, compared to the several hundred that her treatment costs. Laila's point is that women are increasingly willing to subject themselves to chemical treatments and to pay exorbitant amounts to achieve a fair complexion.

One reason for its popularity is that the Indian ideal equation of fairness and beauty is also a European ideal. The name of Laila's salon, 'My Fair Lady' is borrowed from a British musical play of the same name (based on the play 'Pygmalion' by George Bernard Shaw). In the story, a young woman of low class is remade into a lady with the help of a college professor. Fairness, upper class and gentility are equated in the story. This is similar to the equation in India and Kerala, making the shop name doubly effective. Thus, for a young Indian woman, fairness bridges the gap between traditional and modern, contributing to the huge success of the fairness industry in Kerala.

## From background to foreground

In the past, the use of facial colouring and make-up was considered inappropriate. This is consistent with the ideal that women should remain in the background. The traditional *sari* was white and the ideal complexion unadorned. As we have seen, dress styles and use of colours is changing in a way that brings women into the foreground. The same can be said for facial make-up and the use of scents. However, due to the pressures on unmarried women to represent the female ideal, consumption of lipsticks, perfumes and other colour products lags behind that of married

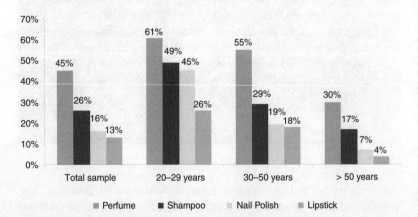

*Figure 3.1* Percentage of women using selected beauty or cosmetic products more than once a week, based on a survey of 408 households in four Trivandrum neighbourhoods

women. Laila told me that her married clients increasingly use lipstick, hair colouring, rouge, eyeliner, and perfumes. This is also true of fragrant body creams, soaps and perfumes. The responses of the female heads of household who participated in the survey indicate an increasing use of these beauty products. Figure 3.1 shows that perfumes are being used regularly by almost half the women surveyed, and that in the 20–29 age group, shampoo and nail polish are also used by almost half the women surveyed. The use of lipstick is still low for the sample as a whole; however, lipstick is used by more than 25 per cent of women in their 20s.

In a study of the consumption of these products in India as a whole, a study by Phookan (2004) showed that 16 per cent Indian women use nail polish at least once a week, and 15 per cent use lipstick at least once a week. Thus, usage of these products in Trivandrum is on a par with usage elsewhere in India. Phookan analyses the market potential of cosmetics and writes that:

> perfumes and fragrances, skin care and hair care products are some of the major segments with promising prospects for U.S. companies... Urban women in the middle and upper income groups in the age range of 23–50 is the target group for international brands, as this group looks for better products and is willing to pay a premium for international quality products...

Since the average Kerala female is married by the age of 23, Phookan's 'target group' in Kerala consists mainly of married women.

Why does marriage mark a change in the ways women consume these products? One explanation is that married women have moved beyond the scrutiny applied to the potential bride. Moreover, I found that husbands who looked for conservative dress and appearance in bridal candidates, often had a changed attitude after marriage. While bridal candidates should look 'natural' and embody tradition, wives can employ elements of the modern in their appearance. Enhancing their complexions with cosmetic products is one way for women to accomplish this. Liechty (2003) found a similar change in the freedoms women are allowed – and take – in the ways they use make-up after marriage.

The first time I talked with Abraham, a 30-year old Syrian Christian, he told me that his wife Sinu hardly ever used cosmetics. I had the feeling that he felt obliged to portray himself and his family as traditional because he imagined that I, as an anthropologist, would be pleased to encounter traditional cultural practices. After we got to know each other better, he gradually revealed his true attitude. It became obvious that he was proud that he had a wife who was skilled in selecting and using new cosmetic products. In one of our conversations in 2003 he told me with a tinge of pride: 'I am sure if she put all of her cosmetics in front of you that you would be astounded. It never stops. There are a hundred different colours.' He also told me that he had begun using cosmetic products himself: aftershave lotion, deodorant and body sprays. According to Phookan's (2004) market analysis, men's bath and shaving products is another of the markets with high growth potential in India.

## Beauty from elsewhere

The growth in interest in beauty and cosmetics is related to massive advertising and to the success of Indian women on the global beauty stage. My experience in assessing the importance of these contests is similar to that of virtually every author in *Beauty Queens on the Global Stage* (Cohen et al. 1996); an initial scepticism followed by a recognition of their importance in style, feminism and consumption. In my research, beauty contests came up in virtually every interview. They were regular subjects of the print media and in television programmes. Market analyst Phookan (2004:5) expressed their importance this way: 'The success of contestants from India at various well known international beauty pageants in the last few years have [sic] also contributed towards making the Indian woman more conscious about looks, beauty, grooming and aware of western cosmetic products/ brands.'

Two landmark events in the internationalisation of Indian beauty occurred in successive years in the mid-1990s. Sushmita Sen was chosen Miss Universe in 1994, and Aishwarya Rai was chosen Miss World in 1995. Their successes in these contests would lead to careers in Indian cinema and centrepieces of myriads of Indian and international beauty advertisements. The winning style of Sen and Rai involved a subtle combination of the traditional Indian beauty (hair, eyes, and complexion) with Western hair styles, lipstick, rouge and slim bodies. Images of women using the same styles can be found in virtually every issue of Kerala and other Indian women's magazines available in Kerala, as well as in newspaper supplements and television advertisements. Beauty operator Laila told me many of her customers ask for a 'beauty queen' hair style, by which they mean the style used by Sen and Rai: shoulder length hair with big curls.

The attention given to these international contests is related to the popularity of local beauty contests, which are regular events in Kerala. They are popular adjuncts to local fairs, and markets, as well as festivals like the Onam celebration. Kerala universities and many localities have their perennial 'Miss'. However, Indian beauty had never been acknowledged on the international scene prior to Sen and Rai's victories. Their status as national icons is reflected in their selection by *India Today* as one of the 25 'newsmakers' of the final quarter of the 20$^{th}$ century. The article summarising their selection celebrates their victories and the recognition of Indian beauty (*India Today* 2000:48):

> Because it felt great to have two gorgeous ladies stand in front of a billion people, keep their nerve, smile, and say the right thing at the right time. Because Indian womanhood is not about looking like the Goodyear blimp wrapped in six yards of fabric... Sushmita's Miss Universe and Aishwarya's Miss World titles was [sic] a coming out parade of in-your-face confidence of such magnitude that men and women's libbers are still in denial. Because nobody cares whether 'Ash' or 'Sush' are fighting or fading – they made it. Period. Because India now owns the beauty business, and it's a refreshing change being the best.

This article captures several telling aspects of the new beauty ideal. One is that clothing should be more revealing of body shape (the body should not be wrapped 'in six yards of fabric'). Another is that plumpness ('the Goodyear Blimp') is out and slimness is in. A psychologist, Rema, who worked with young women claimed in an interview that many middle class teenage girls 'are on starvation diets...The doctors don't know it, but they are going through the starvation thing.' Bina, a young Hindu Nair whose family is searching for a husband for her, told me, 'Nowadays the international beauty is like that: thin, long hair, fair, blue eyes. Me, I am also trying to be a little bit thin.' Her friend and neighbour Rija, a Hindu Ezhava who is a bridal candidate said, 'My friends like to diet, but I can't diet (laughs)! I like food! And our grandmother will not allow us to. She is always telling us to eat something.' The grandmother sees no contradiction between plumpness and beauty, reflecting a generational change in ideas about body and shape.

This change resembles a change in European and North American ideals about body and shape. For example, Counihan (1999) discusses changes in the ideal bodily shape of Italian women. She found that until recently, Tuscan women had an

> active and agentive definition of the body which emphasized doing – working, having children, making food, wearing beautiful clothes – rather than having a perfectly sculpted body...However, Florentine women also struggled with increasingly prevalent images in the media and fashion that proclaimed thinness, female objectification, and beliefs that women should cultivate their body to please men.

One of Counihan's informants told her 'There's a lot of talk about being thin in this culture. If you open the newspapers, the magazines, they are all about shape and diet. They are full of the little image of the little lean woman. When you look in the mirror, you see her and you say to yourself, "Mamma mia, how fat I am".'

An article in the Trivandrum edition of the newspaper *The Hindu* (2001b) reports on a 'mania' for slimness among young women and relates it to the 'beauty queen syndrome':

> According to some, the 'plumping-for shape'[5] and 'beauty mania' was triggered off by the 'Sushmita Sen-Aiswarya Rai' syndrome ...the spin-off of which has been a proliferation of beauty shows... Gone are the days when actresses were hardly exercised over width (weight) as long as they sported smashing looks. This is the era of the lean looking heroine, with a bevy of coy starlets occupying lesser and lesser space on the big screen...The 'stay-as-slim-as-possible' influence of the stars is having, shall we say, a bodily impact on the new generation of girls who are, these days, only too willing to gruel it out to stay as lean as they possibly can. More and more advertising space (both print and air) is being gobbled up by claims of wonder drugs that can 'halve a person' in a matter of a few days...Nutritionists say that as a result the nutrition status of the girls sharply declined to near-anaemic conditions.

Rema was certain that beauty contests and the popularity of international winners in films and advertising had raised the importance of female beauty in Kerala and changed its meaning for young women. She told me:

> We didn't know we were beautiful. When we were growing up we never thought of beauty. Never, never. Now the message is that you

ought to be beautiful and that being beautiful is synonymous with being happy. And they (young women) compare themselves with this (beauty queen) ideal and find themselves short.

A point made by Cohen, Wilk and Stoltje (1996:7) coincides with Rema's insight. They write that beauty contests promote 'the illusion that there is, in fact, a beauty standard, that beauty can be measured objectively, and that beauty has a concrete existence apart from the individual'. An examination of recent studies on gender and fashion in Asia leaves little doubt that international beauty contests are widely popular and that they are affecting 'beauty standards' across Asia (see the contributions of Sen and Stivens 1998). As Peltzer (1993), cited in Fahey (1999:231), wrote about Vietnam, the popularity of beauty contests signals to the international business community 'that Vietnam is open for (the beauty) business'. Multinationals which produce beauty products have initiated and financially supported many beauty contests in India.[6] Both the Miss Universe and Miss World pageants are 'dominated by major corporate sponsorship arrangements with close ties to media empires

**Plate 2**  University students

(Cohen, Wilk and Stoltje 1996:5)', including the principle sponsor, the cosmetic and soap multinational Proctor & Gamble.

These beauty multinationals are part of what Wilk (1996:232) calls 'the global culture industry.' Its marketing power makes beauty 'another consumer commodity... we have the freedom to express our differences, but only with the codes and objects the industry provides'.[7] Fair complexion, slim body, shampooed hair and colour-enhanced lips are associated with new beauty codes and a growing number of products are being offered to help achieve them. Abu Lughod (1990), studying the growing consumption of stockings, make up and shorter dresses among young Egyptian Bedouin women, observes that new forms for beauty consumption lead the consumer into a new source of discipline. She writes that young women 'are resisting their elders by taking up new patterns of consumption (stockings, make-up, and shorter dresses) but are simultaneously being caught in new forms for self discipline'.

There is no doubt that beauty commodification, advertising and beauty contests have influenced changes in the ideal shape of the female body and in the ways that colour, cosmetics and hair styles are used to create

**Plate 3** Syrian Christian woman (married) in *churidar and kurta* (note use of lipstick and *bindi*)

beauty in Kerala. Nonetheless, Indian features also remain important and integrated to the look, including the use of the *bindi* and Indian jewellery. In the past, the *bindi*, a mark or streak of red colour on the forehead, was simply used to indicate that the wearer is married. Today, the *bindi* is used as a beauty-enhancing cosmetic, not only among Hindu women, but also by Christian women (see Plate 3). The new Indian beauty look has been successful precisely because it operates on the cusp between traditional and modern, embodying Indianness and the latest trends from the world's fashion capitals. The commoditisation of beauty is successful because it captures what Mazzarella (2003:203) calls 'close distance', successfully combining aspects of the 'local' and the exotic 'other' from the world beyond India. 'Close distance' works very well in encapsulating what Malayalee want from new products, a point that will be developed in Chapter 9. It has evidently been used successively in China. Hooper examined images of femininity in Chinese advertising of the 1990s and found ample use of the metaphor 'modern flower vase' to encapsulate Mazzarella's close-distance. 'Women should retain their roles as virtuous wives and good mothers, which are likened to the functional roles of the vase, but should be "adorned and bejewelled" in ways that draw on Western ideas of femininity (1998:167).'

In Kerala, India and other parts of Asia, another conduit of beauty ideas from elsewhere is work migration. Female work migrants, or the children of migrants, bring with them new ideas when they return to study or when their parents retire to their Kerala home. Work migration and consumption will be the subject of Chapter 6. Suffice it to say here that family migration was extensive in the 1960s and 70s, and many young women who grew up abroad are returning to Kerala to study or work. Talking with these young women, many of them find adjusting to intense social scrutiny to be difficult. Sangeetha, a 24-year old Hindu Nair framed the dilemma this way:

> You are a Malayalee, you go out, you live there, you are exposed to certain things and you come back. And you don't want to be stuffed back into the shell...living outside, I saw it in some book, is like a rubber band that gets stretched. My mind has been stretched to certain limits because of the various things around me there, the environment, the freedom, and to come back to Kerala, even if you are going to live there, it is kind of sad.

Johnson (1998:232), who focused on women migrants from the Philippines, found that they expressed a similar sense of frustration on returning home. He writes that 'the freedom that they saw as implicit in work abroad was contrasted to a situation in which, according to local protocols, they were supposed to stay at home, to mind the children, not to dress up or wear make-up outside the home or go out of the house without permission of their husbands.' Many of these women challenged these restrictions by 'smoking cigarettes, wearing bright make-up or trying to look glamorous' (1998:232).

## Constraints, ambiguities and changes in small steps

*Swabhaavam* provides friction to changes in the consumption of new appearance and beauty products for unmarried women. It may seem odd to begin a study of changing consumption with a discussion of constraints. However, I agree with Wilk that an understanding of constraints is essential to a full understanding of change. Wilk wrote that in consumption research, more attention was needed 'on limits, and on the institutions, impulses, understandings, and meanings that enforce or sustain them (1999:6)'. As will be explored in the next chapter, a different set of constraints on wives has an accelerating effect on consumption of household appliances. Demands of housework, family care and work outside the home, with little help from husbands and diminishing access to help from other women in the family are some of the important factors behind the purchase of time-saving appliances. Married women's consumption of beauty and cosmetics is seen as a release from these pressures. As Freeman found in Barbados, Kerala women see 'fashion is a liberatory [sic] device transporting women out of the shackles of domesticity or other forms of patriarchal authority...'.

Men are as cognizant of new beauty ideals as women, and are somewhat ambiguous about how they want their wives to dress. Women must pay attention to their long-standing role as the repository of traditional Indianness. They must be the 'picture of cosmological authentication and encompassment', similar to what Johnson (1998) found about women in the Philippines. But both men and women are interested in new beauty products and in the changes borne by international beauty discourses. Consumption of beauty is changing gradually as women combine and mix looks in ways that draw on both traditional and modern styles.

# 4
# The Modern Housewife

Once married, the Malayalee bride is released from the pressures of partner search. However, as a new wife she encounters another set of pressures associated with housework and family care. These pressures and the ways that they are implicated in consumption will be the subject of this chapter. This is not to say that the pressures of dowry and family surveillance are left behind. Many young women are the focal point for complaints from their husband's family and in some cases retribution when dowry payments are not forthcoming or transfers are delayed. Also, most newly married couples of all religions spend time in the husband's family home before moving out and establishing their own nuclear household. During this period, the new bride is given a heavy share of housework and is expected learn her new family's traditions, including those for cooking and cleaning. After establishing their own nuclear household, the responsibility for housework falls entirely on the wife. This chapter will explore the ways that this gendered responsibility affects consumption. I begin the chapter by describing a typical day in the home of a typical middle class, Nair household, headed by Chavita and Anuj, highlighting the ways they accomplish housework.

## A day in the life of Chavita and Anuj

Chavita is an elementary school teacher and her husband Anuj is a civil servant. They belong to the Hindu Nair caste. They live in a typical middle class Hindu home in the neighbourhood called Kannamoola (see Map 2). Chavita is 40 and Anuj is 47. They have two children, a girl aged nine and a boy aged three. Chavita's mother and father, both in their 70s, have a room in the house. Chavita takes the main responsibility for her

**Plate 4** Hindu Nair family, dressed for public. Man in shirt and pants, woman in *sari*

parents, but they also spend time in the homes of her siblings. In many families, the retired parents move regularly between the homes of their married children, where a room or at least a bed is set aside for them.[1] In Chavita's case, her mother is senile, uses crutches and therefore demands a considerable amount of Chavita's time and attention. Until recently, she was assisted by a domestic helper, but her work for the family was terminated as part of an effort to save money.

On this day, Chavita arises as usual at 0530, an hour before her husband. She immediately puts on tea so that it will be ready by the time her husband gets up. She bathes, dresses and then sweeps the hard-packed earth of the garden with the typical Kerala broom, made from palm fronds. She goes out to the street to buy vegetables from a vender who passes through the neighbourhood every morning. Normally, she would have bought vegetables at the market the evening before, then would have cut and washed them, so that she could immediately set to work cooking after she woke up in the morning. However, on this occasion she had not had time to make a trip to the market the day before.

She begins to wash and cut the vegetables and puts rice on the *chula* (wood burning stove) in her back kitchen, which is attached to the house but has a thatched roof. The *chula* is used to prepare special meals for guests and family visits. In many households, it is still used to prepare the daily rice. Chavita and Anuj share the widely held belief that *chula*-cooked rice tastes better than rice cooked on a kerosene or gas-burning stove. Chavita prepares rice on the *chula* almost every day, even though this requires extra work and attention, since the flame must be kept at the proper level for simmering, requiring frequent surveillance and adjusting of the flame. On this morning, Chavita returns to the *chula* frequently to check the flame and to add fuel (mainly coconut rinds).

At 0630 the rice is finished. Chavita places the cooked rice in the electric rice warmer in her front kitchen, where it will remain throughout much of the day. The rice warmer is one of her modern household appliances. She also has a cooking stove, which uses either bottled butane as fuel, a refrigerator and a 'mixie' (electric mixmaster) which she uses to grind the spices for Kerala curries.

Chavita begins chopping and grinding vegetables for curries. Anuj comes into the kitchen, drinks his tea, and helps her to cut up onions. Chavita also begins the preparation of *idyli* (rice cakes) for steaming. She has a big pot and she is proud that she is able to steam 20 *idyli* at a time. While Chavita prepares food, her son moves in and out of the kitchen. She gives him a glass of milk and then begins scraping coconut, an important component that will be used in several of the curries. She grinds the coconut fruit in the mixie and starts a curry simmering on the butane stove. While she cooks she washes dishes. Her daughter comes in to help dry dishes and put them away.

By about 0730 Chavita has finished cooking. She makes lunch packets for herself, her husband and her daughter. She then bathes the children, brushes their teeth, ties her daughter's hair and helps the children to dress. When the boy is dressed, Anuj sits with him in the living room for a few minutes, listening to music on the radio. Chavita then gives her attention to her mother. She helps her mother out of bed (over her protests) and takes her to the toilet. The mother's resistance makes assisting her a heavy job.

At 0800 she gives breakfast to her mother and father and shortly after that to Anuj and the children. She does not have time to eat, so she packs some food to take to work, where she will usually eat her first meal of the day during her break between classes at 1000. At 0820 she finishes dressing and combing her hair. At 0830 she runs out to catch the bus. Anuj leaves for work shortly after she does.

Chavita's son remains at home, under the supervision of his grandfather. Chavita's brother and sister make their usual visit to the house at lunchtime to check on their parents and nephew. The sister serves the lunch that Chavita had prepared before she left for work. The brother helps his mother use the toilet.

Chavita returns home at 1615. She goes to the kitchen and washes the remaining breakfast and lunch dishes. At 1630 four children arrive for their daily tutorial session. Chavita supplements her income by tutoring, which is quite common for school teachers in Kerala. There is a heavy demand for tutoring in Kerala, due to competition for good grades and university admissions. Chavita sits the students down and gives them exercises to work on. She returns occasionally to check on their work. At 1700 she takes a bath, telling her daughter to let her know if the students need anything. When she finishes her bath she sends her daughter to buy milk. She makes tea and begins cooking dinner.

At 1730 she starts rice porridge on the butane stove. Then she goes behind the house to wash clothes at the outside water tap. She washes them in soapy water, beats them on the concrete steps, rinses them and hangs them up to dry. She takes occasional breaks to check on the porridge and to work with her students. Her husband arrives home at 1745 with milk and some cakes from the local bakery. He takes a bath. At 1800, the father of one of the students comes to pick up his child. The other children continue with their studies.

The porridge is ready at 1800. Chavita feeds her son in the typical Malayalee fashion for children under five or six years old, placing food into his mouth with her right hand (in Kerala the right hand is used for food handling and eating). While feeding her son, she puts beans on the butane stove for yet another curry. After they begin simmering, she sweeps the house. When she has finished in the house, she sweeps the garden.

When she reenters the house, her father tells her that it is time to get the lamp ready for evening prayers. She cleans the prayer room, arranges flowers and adds kerosene to the lamp. Then she goes to the kitchen and starts scraping coconut and preparing new curries for dinner while her husband plays with the children.

At 1835 she goes to the prayer room to light the lamp for prayers, but finds that Anuj has already done so and has begun his daily prayers. She waits until he has finished and then she and her daughter pray. Her son, who says he is not feeling well, lies down and rests while they pray. While Chavita and her daughter are praying, Anuj puts away toys and straightens the house. He then turns on the television at 1850 and watches the Malayalee news. When Chavita has finished praying she joins Anuj in front of the television and watches the remainder of the news.

At 1900, the parents of the remaining students pick up their children. After they have left, Chavita, Anuj and the children, return to the living room and sit down to watch a Malayalee television serial (soap opera).

When the programme is over at 1930, Chavita returns to the kitchen. She grinds rice for Achappam or *appam* (rice flour pancakes mixed with sugar and coconut milk). She prepares a vegetable soup for her children.

At 1950 her nephew stops buy. He and Anuj sit in the front room. Chavita prepares and serves them *chai* (tea, milk, sugar and spices) along with cake and banana. She sits with them for a few minutes but does not eat.

At 2015 her brother-in-law (husband's brother) arrives for a brief visit to ask whether the two families should travel together to a wedding reception they have all been invited to the next day. After a brief discussion, the men agree that the families will meet the next day; then the nephew and brother-in-law depart together at 2025.

At 2030 Chavita gives a supper of rice porridge to her mother and father. After they have finished, she serves her husband and daughter the curries she has prepared, along with rice and *appam*. After they have been served she sits down to eat.

At 2115 she clears the table and goes to the kitchen. She washes dishes while Anuj watches television with the children. At 2200 she helps her mother to bed and then puts her children to bed. At 2230 she begins cutting vegetables for the next morning's curries. Anuj turns off the television and goes to bed at 2250. Chavita follows him at 2300, but is awakened shortly afterwards by her son, who is feverish and says he is hungry. She boils milk for him, helps him drink and goes back to bed at 2330.

As this recounting of Chavita's day reflects, her day is filled with tasks related to family care and housework. She gets some help from her daughter and only a little support from Anuj.

## Gender sharing of tasks in the home

I found that the sharing of housework and family care in Chavita and Anuj's household is fairly typical of that in families of all religions, castes and socio-economic situations in middle class Trivandrum. The results of the survey and diaries on gender sharing of housework

*Table 4.1* Participation of male and female heads of households in selected home chores and activities, based on a survey of 408 households in four Trivandrum neighbourhoods

| Chore or activity | Always or often participates (%) | | Rarely or never participates (%) | |
| --- | --- | --- | --- | --- |
| | Male | Female | Male | Female |
| Food preparation | 10 | 88 | 82 | 7 |
| Dishwashing | 23 | 73 | 76 | 7 |
| Clothes washing | 10 | 68 | 80 | 22 |
| House cleaning | 11 | 68 | 81 | 22 |

support this. Table 4.1 shows the participation of males and females in four household tasks.

It is clear from these responses that men seldom participate in household chores. The greatest difference between women and men is in food preparation. This is done almost exclusively by the wife and other women of the household. Men's participation in clothes washing and house cleaning is also minimal. The wife gets some relief from other adult females in joint family households, or in those households which have a domestic helper. The survey responses reveal a considerable amount of sharing in cases in which there are more than one adult female in the household. The second female participated 'always or often' between 40 and 50 per cent of the time in food preparation, dishwashing, clothes washing and house cleaning.

The results from the daily diaries kept by 20 families give further evidence of strong gendering of household chores.

Table 4.2 shows that women used more than five hours a day on these tasks alone. The diaries underline the almost total lack of participation

*Table 4.2* Amount of time used daily on selected household chores and activities by females and males in the household, based on five days of diaries kept by all adult members of 20 households

| Chore or activity | Time used by female(s) (minutes) | Number of cases in which males participate | In those households in which male participates, time used (minutes) |
| --- | --- | --- | --- |
| Food preparation | 208 | 0 (0 per cent) | 0 |
| Dishwashing | 66 | 1 (5 per cent) | 10 |
| Clothes washing | 51 | 3 (15 per cent) | 42 |

from men. The validity of these results is supported a study by of household appliance use in Trivandrum by Vijayakumar and Chattopadhyay (1999). They surveyed 1000 households and found very low participation of males. In fact in only 10 per cent of the households had men ever used any household appliance (cooking appliances, irons, vacuum cleaners, washing machines).

Do wives who work outside the home get more help from their husbands? Anuj contributed a few minutes to dishwashing and straightening the house, but nothing more. The survey results confirm that the low level of participation by men is fairly usual even in cases when both husband and wife work.

A comparison of Tables 4.1 and 4.3 shows little difference in men's participation in household chores in homes where the wife is working. There is evidence from elsewhere in Kerala that working women get little support from their husbands in housework. Devi studied housework in Thrissur and found that 'paid work outside the home does not reduce the family burden and responsibilities of women (Devi 2002:57)'.

Table 4.3 shows that women who work spend less time on clothes washing and dishwashing than do women who do not work outside the home. This is mainly due to the use of domestic helpers. 47 per cent of households in which the wife works have a domestic helper, compared to 37 per cent of households in which the wife does not have salaried work.

Whether or not she has domestic help, the Kerala wife uses many hours a day on housework. The acquisition and use of household appliances provides one way out of the housework bind. Figure 4.1 illustrates the ownership of the convenience appliances, showing differences in

*Table 4.3* Participation of male and female heads of households in selected home chores and activities in households in which women work full time outside the home, based on survey of 408 households in four Trivandrum neighbourhoods

| Chore or activity | Always or often participates (%) | | Rarely or never participates (%) | |
|---|---|---|---|---|
| | Male | Female | Male | Female |
| Food preparation | 9 | 86 | 83 | 10 |
| Dishwashing | 7 | 64 | 87 | 21 |
| Clothes washing | 11 | 60 | 79 | 25 |
| House cleaning | 12 | 60 | 80 | 27 |

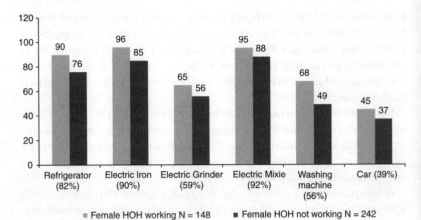

*Figure 4.1* Differences in ownership of convenience appliances between households with female heads of household working outside the home and households with female heads of household not working outside the home (based on a survey of 408 households in four neighbourhoods in Trivandrum)

ownership between families according to whether the wife works or not. Families with wives working full time outside the home are more likely to have all of the convenience appliances.

The ways that appliances are acquired and used in food preparation and clothes washing will be fleshed out below. First, I will explore the reasons for these solidly anchored gendered roles in housework.

## The sources of gender ideology

For Hindus, the ideological references for gender roles in the family are the Vedic texts, the Upanishads and important Hindu sagas, such as the *Mahabharata* and the *Ramayana*.[2] The *Mahabharata* is a source of metaphor on virtually every aspect of life, but especially on the topics of duty and family. The *Ramayana*, through its heroine Sita, is an important source of ideas on the roles and duties of women. In the saga, Rama's wife Sita is kidnapped by a foreign king, held captive, and finally rescued after many years. On her return to Rama's household, he expresses doubts about Sita's loyalty to him and about her chastity during captivity. Sita is forced to prove her innocence by a trial by fire. She withstands the flames of a bonfire, thus demonstrating that she is innocent of Rama's suspicions. According to Chakravarty's (2003:365) interpretation, this episode and Sita's character made the concept of *pativrata* – chastity – the most effective means by which women could become complicit in their own subordination. Chakravarty writes that the *Ramayana* was crafted with 'a patriarchal ideological coherence' and 'Once the norm was in place, women aspired to be like Sita, even if they were required to obliterate themselves in the process (2003:365).'

The message that comes through in the *Ramayana* and other sagas is that in matters of family, women should serve husbands and nurture family. Wadley (1995:117), reviewing these and other Hindu representations of women, writes that the ideal Hindu wife should also be 'good, benevolent, dutiful, and controlled'. There was much debate in the early women's reform movement as to whether the presence of goddesses in Hinduism was evidence of an acknowledgement of female power.[3] This was refuted by early feminists. A 19th century reformist, K. Das wrote 'Whenever there is talk about righting the wrongs of Hindu women, and establishing their equality with men, many refer to the scriptures and speak of their status as the Goddess. But where do we see respect (manifested in everyday life)? (quoted in Bannerji (1991:55).'

Some 20th century feminists have claimed that the goddess Dhurga and her incarnation Kali offer women a source of empowerment (Gupta 1991; Robinson 1999), but Fuller points out how the festivals connected to Kali emphasise power through duty and acquiescence to the female ideal. C. J. Fuller studied the Kali celebrations and concluded that 'Kerala women *could* ... work out their own traumas through identification with Kali,' implying, however that they do not (1992:262). Fuller writes that Kali is 'ferocious' in her unmarried form; however, wifely goddesses – Lakshmi, Parvati or Saraswati – all represent 'a subordinate power lacking powers separate from her husband's (1992:41)'. A study by Usha (2004) in 2003 asked women in 200 households who embodied their womanly ideal. The majority answered the goddess Parvati (Shiva's consort). Usha encapsulated the reasons in the following way: 'She (Parvati) was the epitome of goodness, beautiful and dutiful, submissive, and yet powerful (2004:27).' This finding supports my own: first, that Hindu ideology is still a source of reference for everyday conduct; and second, that women's empowerment is related to the fulfilment of her duties to husband and family.

From the early 20th century, modernist and feminist reformers aimed at revising the female ideal. They attacked submissiveness, associating the modern woman with qualities such as assertiveness, orderliness, thrift and intelligence (Jain and Mahan 1996). Nonetheless, women should retain their traits of 'chastity, obedience, self-sacrifice, adaptability, modesty, nurturance, domesticity, and being "home-loving" ...'; and should not forget 'the importance of social appearance, and social conformity...while discouraging independence or pursuit of individual goals (Mukhopadhyay and Seymore 1994:4)'. The qualities of thrift and intelligence were reinforced in educational reforms for women in the early 20th century. Home management was added as a favoured subject for women's education (Bannerji 1991). Women reformists of the time supported these changes. Reformer K. Das (cited in Bannerji 1991:58) wrote, 'Running a home requires controlling family affairs, maintaining order and accuracy, hard work, thrift, frugality, skill and judgment, all of which will be improved with education.'

The Indian woman's reform movement was influenced by reforms in Europe and North America. The early reforms there were also focused on home management and development of skills such as thrift and maintaining good order. Henrietta Moore writes that one of the best known of the early European reformers, Catherine Beecher 'accepted a conventional definition of the domestic world as women's sphere, but she argued that women should rule the home in their capacity as skilled professionals... (1994:84)'. Educational reforms for women emphasised home management. Wives should be educated to be efficient household managers.[4] According to Devika (2003b), husbands were accorded the paternalistic status as guides on their wife's 'journey to modern personhood'.

The first priority of Kerala women is to take care of house and family; this comes before employment outside the home. This applies to both Hindu and Christian women.[5] A Syrian Christian, Abraham, told me, 'It doesn't matter what she is. She could be a Harvard student. You see Mea (his wife, who has a Master's Degree) now even as homemaker. It is the perception that it is the man of the house who has to feed the family.' In many cases, this 'perception' becomes the rule. Many husbands do not want their wives to work. This includes highly educated men. Murickah (2002) studied educated, urban married couples of all religions and castes in Kerala. She found that 55 per cent of the men stated that they did not want their wives to work after marriage. Only 9 per cent of the husbands and 4 per cent of the wives preferred that the wife be the one to decide whether she works or not.

Fewer Kerala women work than in the majority of other Indian States, in spite of higher educational achievements. In 1991, Kerala ranked first amongst Indian states in female educational achievement, but ranked only 22nd (of 27 states) in women's participation in the workforce (Census of India 1991). In 2001, women made up only 24 per cent of the Kerala work force, compared to 32 per cent for India as a whole (Census of India 2001). Among the survey participants in Trivandrum, 96 per cent of males with high school degrees are employed compared to 38 per cent of women. Furthermore, every male with a university degree are employed compared to only 34 per cent of females.

Halliburton (1998) proposes that, ironically and even tragically, education adds to women's frustration with the housewife identity; tragically, because he hypothesised a relationship between this 'bind' and Kerala's high female suicide rate. Mijeong Lee (1998) found a similar frustration in her study of South Korean educated women. Lee writes that the South Korean government has long emphasised gender neutral educational opportunity, but she found that the role definitions of the highly patriarchal South Korean family continue to impose a role in which women are to serve husband, children, and in-law family. 'Women ... should neither choose nor judge what they do...Educated women are often frustrated by not being able to use their education in a productive way (1998:169).'

I found that among men who were willing to allow their wives to work, most reserved approval of the type of work. The favoured kinds

**Plate 5**  Hindu Ezhava family, dressed for home. Man in shirt and *lungi*, woman in 'house dress'

of work for both Hindu Nair and Ezhava women are as teachers and in government jobs. Many men are sceptical to their wives working in commercial workplaces. Hindu Ezhava Sarita (24 years old), who had a Bachelor's degree but who had struggled for four years to find a teaching job, decided to obtain a licence to drive an auto-rickshaw. Her husband was adamantly opposed to this.[6] He felt it was not a dignified profession and forbade her to work even though he was unemployed and they sorely needed the income.

To summarise this section, it is women who do housework. The ideal woman is both caring and efficient. Given the declining use of domestic help, convenience appliances are attractive, especially in households in which the wife is employed. The use of appliances like washing machines, mixmasters (mixies) and microwave ovens promises to reduce the time women spend on washing and cooking.

I now take a closer look at how gender practices, time pressures and new technologies are implicated in changing consumption of cleanliness and food.

## Clothes washing

Women who do not have a washing machine wash clothes by hand. This is a time consuming practice involving hard physical work. Clothing must be soaped, beaten to remove dirt particles, rinsed, wrung out and then laid out (or hung up) to dry. The acquisition of a washing machine bears the promise of the relief from hard manual work and the freeing up of an hour or so a day of time (the diaries showed that women use on average about an hour a day for clothes washing). The time saving potential is an important motivator, and this shows up in the differences in ownership of washing machines between families in which wives work outside the home. Sixty-eight per cent of households in which women are working have a washing machine, compared to 49 per cent of households in which the wife is not working. Since families with two incomes have greater purchasing power, this might explain part of the difference in ownership (mean income in households with working wives was about 23,000 rupees per month, compared to 13,000 per month for households with non-working wives); however, for entertainment appliances such as televisions and comfort appliances such as air conditioners, there are no significant differences in ownership of these in families with and without wives in salaried jobs. This supports an interpretation that time considerations are behind the rapidly increasing demand for washing machines; based on sales data provided by Trivandrum retailers, the sale of washing machines increased by 500 per cent from 1991 to 1999 and by 30 per cent yearly from 1999 to 2001.

The story of Ajay and Sinu's (Syrian Christian) purchase of a washing machine gives insights into why people want washing machines. Ajay and Sinu are married and have a three-year old son, Levis. Both Ajay and Sinu have full-time jobs. Over a period of six months they fretted about how to lighten their housework and child care duties, given their long work days, including Saturdays. They considered putting Levis in a kindergarten, but had calculated that the costs would strain their household budget to the maximum. They decided that hiring a domestic helper would be a less expensive solution. For months, they looked, unsuccessfully, for a domestic helper from the area near their home. They finally found a woman from a village about a one-hour bus ride from Trivandrum. She was given the tasks of caring for Levis and doing the household chores: cooking, washing clothes and sweeping the house. However, because of the unreliability of the bus system, she was frequently delayed or absent. Also, the helper complained that the combination of housework and child care was too difficult to manage. As a result, Sinu decided to let her concentrate on two tasks, taking care of Levis and cooking, which left the cleaning and washing to Sinu. The only time available for Sinu to do those chores was in the late evenings after she came home from work, or on Sundays, her only day off.

After a long period of vacillation, including calculating and recalculating household income and expenses, Ajay and Sinu finally decided that the expense of a washing machine was justified. After many weeks of shopping, they found an appliance retailer who offered them the following deal: a low down payment (one-third of purchase price) and low-interest instalments over a three-year period. This type of financing dates from the mid-1990s, but is rapidly becoming standard retail practice in the appliance business. Ajay calculated that this form of payment would make it possible for them to manage the monthly payments; so, they bought the washing machine. This story shows that the time-saving and time-reorganising potential of the washing machine are important factors in changing consumption.

## Food preparation

As we have seen, food preparation is another time consuming activity which is the exclusive domain of the wife. Convenience appliances associated with cooking, including refrigerators, mixies and microwave ovens offer the prospect of saving time and work. Examining Chavita's day shows that food preparation and cooking seem to be a never-ending activity. The time demands of food preparation were also revealed in diaries of women in the sample of 20 families who maintained daily activity diaries. Women used on average over three hours daily preparing food. In large families, women used up to six hours daily preparing food. Much of this time involved grinding the spices and condiments used in staple dishes like curries and *sambars*. Grinding, cutting, scraping and crushing spices and condiments take time. Electric mixies and grinders provide a way to significantly reduce the time devoted to these tasks. Figure 4.1 shows that 92 per cent of households surveyed owned mixies. In rural areas around Trivandrum where houses had recently connected to the electricity grid, interviews with housewives revealed that the 'mixie' was among the first appliances purchased (along with lamps, electric irons and televisions).

The mixie is popular because it can be used for tasks that cannot be delegated to domestic servants. The helper often cuts up vegetables, fish or meats for cooking, then the wife or other family women do the cooking, which starts with the mixing and grinding of herbs and spices and continues with boiling, frying, steaming and simmering.[7] Even though members of the Nair caste take work as domestic helpers, the division of tasks between wives and helpers in cooking has its roots in ideas about caste purity. Gupta (2000:35) writes that Hindu caste ideology involved a view that 'people of different castes are naturally different because they are made up of different substances. Though these substances are invisible to the naked eye, they can invade other beings through the various orifices in the body'. Even the touching by lower caste members of vessels in which food was to be prepared could lead to 'strong pollution' (Dumont 1980). No one today talks about division of work in the house in terms of caste or purity, but in very few middle class homes does the domestic helper participate in the chopping and grinding herbs and spices for curries.[8] Since these tasks are the exclusive task of the wife, the purchase of a mixie saves *her* time. This contributes to interest in the mixie.

Another widespread attitude about food preparation affects time and the ways people consume household appliances: once prepared, food should be eaten the same day, and should not be stored for reheating later. This affects attitudes to bakery and take-out foods, because it is not transparent to the buyer how much time has elapsed since they were prepared. Our Nair neighbour Anil would wring his hands in consternation every time I stopped at a bakery and bought a *somosa*, vegetable roll, or other prepared food. His wife Deeba did not agree with Anil's aversion to fast food. Deeba did much of the cooking for Anil's joint-family household. Since she worked as a teacher, her hours at home were filled with cooking for the seven members of the household. In order to save time, Deeba began to occasionally buy and serve ready-cooked foods from the local bakery. This offended Anil's mother and aunt, who were the senior women in the family. There were a series of conflicts over food preparation, and according to Anil, this was an important reason why he and his nuclear family moved out of his father's home.

Attitudes towards storing and reheating foods have affected the purchase and use of refrigerators. Refrigerators were introduced in the 1960s. Manufacturers expected the same rush to buy them in India as had been the experience in Japan a decade earlier. This rush did not happen in India. Forty years after they became widely available, only 60 per cent of middle class families in Trivandrum owned a refrigerator (based on the survey of middle class homes in 2002). Refrigerator purchases ranked far behind the purchase of televisions (95 per cent ownership), and only slightly ahead of much more expensive cars (46 per cent ownership). Gopal, a Hindu Ezhava who had spent much of his life in Kuwait, found it puzzling that Malayalees were not exploiting the full convenience potential of the refrigerator.

> The refrigerator, the tradition over here is that you prepare food for the day and you try to finish it or you throw it out. So there is a lot of wastage. We in the Gulf, we preserve. Both my parents were working and they made the food for a week and we utilised it each day. They used to cook on a Friday which was a holiday for them. She (his mother) would take a part for the daily use and then warm it up. So there it was a necessity so we continued that necessity back here. But my wife finds it difficult to eat this. Even if they (the typical Malayalee) have a refrigerator in their house it is not put to much use. They don't think it is a great necessity. So it is a luxury for them and for us it is a great necessity.

Gopal's introduction to a different way of preparing and consuming food, and the role of the refrigerator in it, is responsible for his changed attitude about food and refrigeration. I will return to work migration and the ways it affects ideas about consumption in Chapter 6. Here I follow up on his observations about 'wastage' and that the refrigerator is 'not put to much use' in Kerala.

The refrigerators potential to save food preparation time by cooking in bulk, storing and reheating is not something most Malayalee households of middle and older generations exploit. The aversion to stored foods coincides with ideas about food storage in ayurvedic medicine: Food should be 'alive' in order to give life to the eater.[9] Raw food is more alive than cooked food. Overcooked, undercooked, burned, bad tasting, unripe or overripe, putrefied, or stale food should never be eaten. Leftovers should be heated up as soon as possible or, ideally, avoided altogether (Sen 2004:170). Eating foods that had been refrigerated and reheated was said to cause digestion problems and laziness. According to Marriott and Inden (1977, cited in Kolenda 1991:62), classic Hindu texts convey the point that 'Processes like digestion and sexual intercourse require heat to separate, to distil and to mix different (bodily) substances.' Another related belief is that cold food and drinks, including refrigerated water cause sore throats, colds and stomach problems. Our housemates and neighbours, who did not usually comment or intervene concerning our food habits, would invariably get agitated when we gave cold water or chilled bottled drinks to our children. We were admonished time and again about the risks of drinking cold beverages or eating ice cream. Ingesting something cold is thought to impede digestion, which in turn leads to sluggishness and laziness.

Despite the reservations of middle-aged and elderly people concerning cold drinks and stored food, we noticed that our younger neighbours kept all sorts of bottled drinks and left over food in their refrigerators. It is clear that there are generational differences in food-storage practices, but it is also clear that ideas are changing more slowly among Hindus. Only 32 per cent of Hindu families regularly drank chilled soft drinks compared to 49 per cent of Christian families.

The theory of Akrich (2000, building on ideas by Latour) on technology agency is useful in interpreting these generational differences in food attitudes and practices. Akrich wrote that technologies such as the refrigerator come embedded with 'a framework of action (2000:208)'. The word framework is meant to imply that technologies are not determinative, in that they do not homogenise practices, but rather that they have a potential to channel practices in certain predetermined ways. The refrigerator's embedded frameworks are full of potentials for differing kinds of uses. While the motive for purchasing the refrigerator may only involve a limited number of these, its other latent potentials may lead to unanticipated new uses. Interviews with middle aged and older people who had purchased a refrigerator revealed that for many of them, the principle motivation was to save space. This might seem strange from a European perspective, but for those without refrigerators in Kerala, raw food products such as vegetables, meats, fish, milk, curd and eggs are usually laid out on shelves in a dark room. The refrigerator eliminates the need for storage space.[10] One of our neighbours had purchased a refrigerator and converted the food storage room into a dining room.

Regardless of the original motivation, once purchased, the potentials of the refrigerator to save cooking time and to save trips to the store have contributed to changing practices. Refrigerator owning younger families who have grown up with refrigerators are much more likely to prepare food in bulk and serve it up after it has been stored for days in the refrigerator or freezer. They are also more likely to store and imbibe cold drinks. This change involves a two-directional, or distributed agency between the refrigerator technology and the changing sociocultural contexts of household practices. Chavita said this about refrigeration and reheating of food: 'These ideas are changing. Now people are working and going to school in the morning. There is no time to prepare every meal.' Women are working and seeking ways to alleviate time pressures. The refrigerator provides a means to save time. This practical urgency is in turn behind changes in food ideology.

The increased interest in refrigerators is related to the increasing use of another convenience appliance, the microwave oven, which was designed to quickly heat, re-heat, and defrost and cook frozen foods. In Trivandrum, 15 per cent of the families surveyed owned a microwave oven. The sales manager at one of Trivandrum's largest retailers of electronic goods told me that sales of microwaves were growing faster than sales of all other household appliances. For India as a whole, sales grew from 5000 units in 1991 to 235,000 units in 2002 (Statistical Abstracts India 2003). The refrigerator, freezer, and microwave constitute a mutually reinforcing technology regime which is highly agentive in changing food consumption.

## Conclusion

At the core of this chapter is the housewife. The responsibilities for housework and child rearing are squarely on the wife's shoulders. This has roots far back in the period of *marumakkathayam*, when women of many castes performed housework for their *taravad*. As patrilocality became more common, and wives co-located with their husbands, women performed housework for their husbands and his relatives. Exclusive responsibility of housewife for housework has not changed despite significant gender and educational reforms. In the meantime, other developments have increased time pressures on the wife. One is the gradual entry of women into the work force, diminishing the number of hours available to them to do housework. Another development affecting middle class and elite families is the increasing difficulty in finding domestic helpers. The number of women willing to be domestic helpers is decreasing while the wages demanded by domestic helpers are increasing. Yet another development is the decline of the joint family household, in which women share chores. These changes have contributed to making convenience an important motive for the purchase of household appliances. As Preethi, the 17-year old Nair told me about her imagined home after marriage, 'I think they (convenience appliances) will help a lot. Before buying all those things was considered like social status, but now they are necessary.'

The designation of household appliances as convenience items, and the marketing of their time-saving features resembles marketing in North America in the 1930s and 1940s (Cowan 1989); in Japan in the 1960s and 1970s (Wilhite *et al.* 1997; Garon 2003); and elsewhere in Asia, notably China, from the 1990s (Hooper 1998). General Electric, the transnational manufacturing giant which pioneered appliance marketing associated 'progress' with their products (their motto: 'progress is our most important product'). Progress, being modern and having modern appliances were all part of the ideal late 20$^{th}$ century wife. This discursive trio (progress, modern, efficient housewife) was also associated with freedom, because owning appliances would free women from housework.

North American feminists began to question this promise in the 1960s. Cowen's (1989) study as well as another similar study by Vanek (cited in Houchschild 2003) both showed that appliances did not save women as much time as promised. Cowan's study compared housework in the United States in 1940 and 1965; Vanek between 1920 and 1960. Both found that despite dramatic increases in the ownership of washing machines and dishwashers in the intervening period, women in the 1960s used about the same amount of time on housework as women at the beginning and mid-century did. The time women used for some tasks declined in the 1960s, but because women still did most of the housework, new tasks materialised that filled the freed-up time. Vanek mentions greater time used in maintaining and repairing appliances, more shopping, washing clothes more often and keeping account of household expenses. In Trivandrum, as long as gendered practices and the ideology that supports them remains firm, the potential exists for the same filling of women's time with new tasks and new appliances.

A review of recent ethnographies from other Asian countries (see Hooper 1998 and Sen and Stivens 1998), from Africa (Hansen 2000) and from the Caribbean (Freeman 2000) reveals some of the same pressures on gender identities, division of household work and time as I found in Kerala. For example, Freeman writes that the myth of the Caribbean 'strong matriarch' is still used to justify assigning women both hard work in salaried jobs outside the home *and* full responsibility for housework and children. She found that 'These competing gender ideologies (hard worker and mother) coexist, not only in the discourse of the state and TNCs, but also among the women themselves (2000:11).' Women configure their lives and use consumption 'in ways that draw upon and reinvent these competing ideals, producing both liberating and constraining effects ... (2000:11)'. Writing about China, Hooper (1998:179) cites MacKinnon's (1989) finding that 'Women become as free as men to work outside the home while men remain free from work within it.' Hooper calls this a 'double burden', which she sees as relevant for gender binds in both socialist and capitalist societies. An important difference between developments in Trivandrum and those in China and elsewhere in India is that women are moving into the workforce at a slower pace in Kerala, which is also attributable to the strong gender ideology that fixes women's place in the home. Nevertheless, whether they are theorised as 'double agency' or 'double burden', new discourses on women and on women's roles in the family press for women to be simultaneously modern and traditional, and to work both inside and outside the home. Consumption is deeply implicated in the efforts women make to negotiate these dualities.

# 5
# Exercising the Extended Family

Bina, a 43-year old Nair woman, her German husband and their two teenaged children recently returned to Kerala after having lived in Germany for 20 years. In talking about contrasts between life in Kerala and Germany, Bina repeatedly emphasised that the most striking difference had to do with family and family relations. She found that in German families like her husband's, everyday life was 'barren of contact'. By contrast, she described her everyday routine in Kerala as full of 'cousins, cousins, cousins, and aunties. I cannot even grasp the huge amount of family contact. My boys (teenagers) are often saying about the visitors, "that is enough"'. It was obvious that she was pleased by her return to the Kerala family. Many Malayalee pointed to family and family relationships as the things that most clearly distinguish Kerala from the West, where family is imagined as weak or even dysfunctional. For those who have not lived in the West, this imagined Western family is undoubtedly influenced by Western movies and television soap operas which play on family feuds, divorces, extramarital affairs and gossip.

The extended family in Trivandrum is important both as a social unit and as a social network. As Connell (1987:121) writes about the Indian family: 'In no other institution are relationships so intensive in contact, so dense in their interweaving of economics, emotion, power and resistance.' This chapter will examine the Kerala extended family, the ways the ideal of family is 'exercised' (in the words of Bourdieu) and how consumption is implicated in the 'practical and symbolic work of reinforcing family (1998:68)'. I will show how the family is a mediator of social and material flows, both within the nuclear family, the extended family and, in the case of marriage, between two extended families. Of all the ways in which family is exercised, marriage and dowry place the most extensive demands on the family's economic and social resources. The discussion in this chapter takes up the examination of dowry and consumption begun in Chapter 3, with a focus on its constituents, its role in household economy and its relationship to consumption.

I begin by depicting relations in two extended families, one a typical 'Gulf family' from the Ezhava caste, the other a Nair family with no family members working outside India. To begin with the latter, Govindan Nair (62), his wife Meenakshi (60) and his sister Rajama (57) share a joint family household. They have three adult children, Anil (41), Meena (34) and Lakshmi (31). Rajama, Meena and Lakshmi are school teachers. Anil works at odd jobs repairing electronic equipment. Anil and his wife, Deeba, and their two sons have their own home, but as is typical of a senior (in this case only) son, Anil spends lots of time in his parent's home. Meena, whose partner search is described below, recently married and lives with her husband in a village about an hour's drive south of Trivandrum. She is a frequent visitor to her parent's home.

In a neighbouring Ezhava joint family, Bala is the family-head (referred as the *karnavar* by the other members of the extended family). Bala is a retired government worker. He and his wife Geetha have three sons and three daughters. His unemployed middle son Akbar (43), with his wife and children, moved into Bala's home after Akbar's job contract with an oil company in the Gulf was terminated. Akbar spends much of his time plying job contacts and looking for new employment in the Gulf. One of Bala's daughters Bina (34) and her two children also reside with the family. Bina's husband works in the Gulf, only returning to be with Bina and the children once or twice a year. Bala's oldest son Jamael lives and works in North India. Two of Jamael's daughters (Bala's granddaughters), Rija (24) and Lakshmi (18) reside temporarily in Bala's joint family home while they complete their university studies in Trivandrum. Bala's youngest son, Barun (29) is unmarried. He has a job contract in the Gulf.

The fathers in both of these families have much to say about major household consumption decisions, such as those involving the purchase of kitchen and entertainment appliances, but also maintain control over the consumption of all of the household members. Rija, Bala's granddaughter, told me that she was not allowed to make any purchase without consulting her grandfather, with the exception of very personal things such as toilet articles and underclothing. In addition to controlling consumption, senior family members control the household economy, to which all working members are expected to contribute. In Bala's family, this includes both of the sons who are working abroad.

The role of family *karnavar* has a long history that was explored in Chapter 2. The *karnavar* in matrilineal households had authority akin to a feudal lord, exercising control over economy, housework, marriage partners and many other aspects of family life. In the 20th century this authority was weakened by the breakup of large joint family households and laws giving rights and authority to husbands. Joint families were rearranged according to patrilineal principles, with the children and kin of the senior men sharing the household. Through this transition, the authority of senior men remained strong and respect was exercised by other family members in a number of ways. Rajima, Govindan's 57-year old sister, was one of many of the elderly generation who talked about the authority of senior men in her childhood household. She said that:

> When we used to sweep the house we wouldn't touch grandfather's chair with the broom. Our grandfather's shoes – we never touched them with our legs or broom. We always took them with our hand when we had to sweep the room. Now the children will use their feet to move them from one place to another place.

Rajima lamented that children no longer show adequate respect for senior family members. Govindan and Bala agreed. Govindan used his own 11 and 13-year old grandchildren as examples. According to Govindan, they neither respected him or their father Anil. 'They don't care for the father-son relation or care for other relatives. Nowadays the development is in this way.' He assigned most of the blame for this on the new generation of mothers, who were not making an effort to discipline children to respect their elders.

Claims about loss of respect for family elders were often couched in issues involving consumption. For example, 24-year old Rija said that while she still consulted senior members of the household about purchases, she often found herself to be 'at opposite poles' when it came to decisions to buy and use things like fast food and cosmetics. She attributed the conflicts to her grandparent's excessive conservatism regarding food and its preparation and to their excessive concern for thrift. According to Rija, for her grandparents every new purchase should be carefully weighed and related to a specific need. By contrast, Rija and her cousins were interested in experimenting with new products. Many young people couched consumption in terms of freedom. Consumption is used as a testing ground for the bonds of tradition, gender and family discipline.

Neither Govindan nor Bala were interested in struggling to maintain their authority over the family economy. In fact they both seemed relieved that their children were establishing their own nuclear households and taking over economic responsibility for themselves. As Bala put it 'Everyone who is married should take (economic) responsibility (for themselves).' Many of the younger generation see the establishment of a nuclear household as a way to break the bonds of family control and to facilitate greater freedom to consume as they wish. One of Bala's daughters, Lakshmi (whose husband works in the Gulf) recently moved out of the joint household and into her own home, despite the increased pressure this put on her economy. She attributed her decision to a desire for greater freedom; freedom to come and go as she pleased and to buy things without having to consult with her parents.

These associations of the small, nuclear family with freedom and autonomy are related to reform-oriented discourses encouraging the nuclear family, parenting and family planning in Kerala from the early 20th century (Devika 2002a and 2002b). These discourses encouraging the small, modern family are not unique to Kerala and India, but have been evident across Asia and Africa (Ferguson 1999; Garon 2003). In Kerala, this discursive remaking of family has also been promoted by social reformers, educationalists, family counsellors, and popular magazines. According to Devika, the reformers claimed that parenting would be more effective and less expensive in smaller families, which would (so the reformers claimed) ultimately be good for both the collective (Kerala) and the individual family. The Kerala government initiated ambitious family planning programmes beginning in the 1950s and carried on aggressively promoting family planning through the 1970s.[1] The declining birth rate, together with the decline of the joint family as a permanent living arrangement, meant that within two or three generations, household sizes were radically reduced.

An examination of the history of family and household might well lead to the deduction that the decline in the joint family household has been responsible for a decline in the strength of the extended family. At the beginning of the 20th century 80 per cent of Kerala households were joint family households. In the course of a half century, this changed dramatically. By 1965, two-thirds of Kerala households were either nuclear households, or a modified nuclear household, consisting of the married couple, their children and lineal (parents or grandparents) relatives (Government of Kerala 1972). In 2002, in the 404 households surveyed in Trivandrum, 94 per cent of households were nuclear households (although 35 per cent of these had one or more members of the parental generation sharing the home). The results of my research are that in spite of the physical separation of extended families into different households, the strength of family and of family ties remains strong. In Bala's daughter Lakshmi's case, in spite of her desire for her own home, she venerates her parents and has close contact with her siblings and their children. She visits her parent's home virtually every day and talks with her mother frequently on the telephone. Rija, commenting on the solidarity of the bonds between Lakshmi and her mother, said this:

> My aunt (Lakshmi) lives in a separate house. If she doesn't come here for two days, my grandma wails, 'she didn't come, she didn't come'. At least once in two days she comes here. And they talk daily on the telephone. My auntie will always bring something with her, for example, banana. She has a vegetable garden. If she is picking things, at the same time she will bring some here also. And if grandmother has something, she (aunt) will take it.

I found that Lakshmi's frequent contact with her parental home and sense of obligation to her parents is typical of adult children. Sibling and cousin relations are also strong. Elder brothers of all castes expressed a sense of responsibility for their sisters, not only for helping them to find husbands, but also for providing economic support after their marriage. Members of the Nair caste attributed this since of caring and responsibility to the legacy of *marumakkathayam*. As discussed in Chapter 2, in matrilineal households, women's relationships with brothers and maternal uncles were more important than relationships with husbands. However, I found strong sibling and cousin relationships in both Hindu Ezhava families and in Christian families, neither of which have this matrilineal legacy. Raheja and Gold (1994) claim that strong brother-sister ties are characteristic of families across India. Thomas, a Latin Christian who has worked his entire adult life in Saudi Arabia, often lamented that he had not accumulated anything from his many years of work in the Gulf because of his constant servicing of the economic demands of his three married sisters.

Today's extended family is spread across the households of its constituent nuclear families. However, the sense of belonging to family is still strong and the exercising of family relationships remains important. One Hindu Nair man, currently living in a nuclear family, described the difference of that experience from the joint family household of his youth as follows:

> In the joint family when we needed help, every one together used to discuss about that. Now I am living in the nuclear family. Even

so, I will discuss with other family members when an important thing is coming. For example when my daughter was going to get married I called every one who is older than me and discussed with them. The only difference is that in a joint family, my parents and sisters *see* my problems then they help; in a nuclear family I have to inform them.

As this man expresses, important events in the lives of family members are the source of discussions, consultations and in many cases the exchange of goods. On the one hand, one can say that the idea of family and the relations within the family are reinforced through these exchanges, and on the other that the flow of goods is one of the important medium of reinforcement. In the following, some of the most important of the events and milestones that draw a family together are discussed. This is followed by a discussion of dowry, which is the source of the single most important flow of goods and money through family.

## Events and celebrations that reinforce family

An important milestone and family celebration for Hindus is the *perideel*, the 'naming ceremony' for babies, which is celebrated 27 days after birth for Hindu boys and 28 days for girls. The family gathers for a ceremony in which a black thread is tied around the female child's waste and bangles are placed on her arms and ankles. For Christian children the naming takes place in conjunction with baptism rites, also attended by members of the extended family. When Hindu babies are six months old they are given solid food for the first time in a ceremony called *chorunu*, the first ritual feeding. A priest or senior family member puts strong tasting foods in the baby's mouth, including garlic, honey, pickles and chilli (see Plate 6). Ethnographers Gough (1962a) and Puthenkalam (1977) noted the importance of both of these childhood rituals in the early 20th century. Puthenkalam wrote 'Though simple, these ceremonies attained a certain grandeur when the majority of the members of the *taravad* attended them (1977:178).' Based on my observations, these childhood rituals remain important ceremonies that draw family members together and celebrate the family.

Another event in which the extended family participates is the 'learning ceremony', practiced by both Hindu and some Christian families. In this ceremony, one of the family children traces its first Malayalam character in a plate of uncooked rice with the help of a priest or senior member of the family. Children's birthdays, which follow the astrological

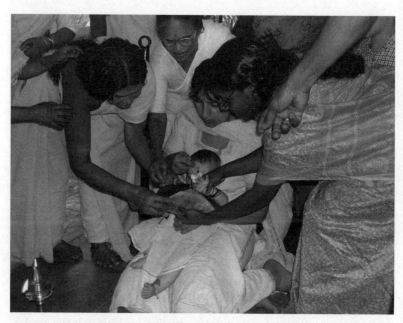

**Plate 6** The *chorunu* (rice feeding) ceremony

lunar-based cycle (and not the Occidental calendar) are also occasions for family celebrations. Marriages are mid-life ceremonies that are very important family events; these will be discussed in some detail below. The death of a family member brings together the entire extended family. In Hindu families, members gather at the house of the deceased, where they remain sequestered for several days after the death.[2] After the cremation the deceased's family members begin a period of fasting, during which they drink only black tea and eat only green gram (thought to cleanse the body). On the third day the deceased's ashes are removed and washed. The urn holding the ashes is placed on an altar in the home of one of the children. A holy lamp (called *Sanjaanam*) is lit over it every day. After a year the ashes are cast into the waters of one of Kerala's holy rivers; this is thought to be the starting point of the soul's journey to the spiritual world. Amongst Christian families, only Jacoban Christians sequester themselves after the death of a relative; however, Christian families of all denominations gather for a wake (feast).

Religious holidays and Kerala's numerous festivals are occasions for drawing the extended family together. Some celebrations are unique to Kerala, some are common to South India and still others are pan-Indian

*Exercising the Extended Family* 75

**Plate 7** The *Vishukanni*

celebrations. *Vishu* is one of the pan-Indian Hindu celebrations which is practiced by the extended family. It takes place in April, when the Hindu calendar places the earth in an auspicious position relative to the sun. On the day of *Vishu*, the first thing a person sees on rising from bed is said to determine that person's fate for the coming year. As family members rise, they cover their eyes and are lead by one of the family's senior members to the *Vishukkani*, an elaborate arrangement of fruit, jewels, coins and flowers placed around the base of a burning lamp. Everyone opens their eyes in unison (see Plate 7). The beautiful sight seals the family's good fortune for the coming year.

For Hindus there is one annual event that engages families over a period of several weeks. The *Sabarimala* pilgrimage has important religious significance (discussed in Chapter 8), but is also an event in which family men bear the family's fortunes with them on a trek to visit the God Ayyappan[3] at his mountain temple and shrine about 200 kilometres from Trivandrum. At least one male in virtually every Hindu household in Trivandrum has made the *Sabarimala* pilgrimage. In the households surveyed, about half of the male heads of household had made the *Sabarimala* pilgrimage in the previous three years. In many families, many of the family men, as well as female children who have not reached puberty and non-menstruating older women, make the trip together. I participated in the Sabarimala pilgrimage with Anil and six of his friends in December 2001. About 20 members of Govindan's extended family (I was considered to be an adopted member) gathered to participate in the departure ceremony. Govindan began with prayers and blessings for the pilgrims. He had prepared a cloth bag containing several coconuts and offerings for Ayyappan (the *irumudi*). In addition, each family member approached the pilgrims in turn and gave them a few coins to support their journey. After the departure ritual was completed, I and my fellow pilgrims made a night-time journey by car, arriving after midnight at the base of the mountain on which the temple is located. After a bath in the Pamba River, we and thousands of other pilgrims began the six kilometre walk up the mountain. Because of the large crowds, only a small number of pilgrims were actually allowed to approach the shrine of Ayyappan and view the idol. We were fortunate to be allowed to approach and offer a prayer. When we returned to Trivandrum the next afternoon, the entire joint family gathered to hear the story of our trip. Govindan was elated that the pilgrims' offerings had been accepted. He told and retold the story to neighbours and acquaintances over the following days and weeks.

From the point of view of consumption, the most important aspect of this ritual is the exercising of the family. Family fortunes are bound together and born by the pilgrim to the shrine. This ritual reinforces the importance of the extended family and that family fortunes may be borne with members across significant geographical distances. Work migration has analogies to this Sabarimala pilgrimage; I will explore in the next chapter how migration involves marshalling of family resources and a family member bearing family fortunes with them as they seek work outside of Kerala. This is reciprocated by a reverse flow back into the Kerala extended family of goods, money and, most importantly, new consumption practices.

In their analysis of the *Sabarimala* experience, Osella and Osella (2003) emphasise the reinforcement of the participant's masculinity and their roles in procreation. Osella and Osella write, 'Men tell us that they go to Sabarimala with children in mind; to protect those already born; to ask for the conception of those desired but yet unborn; to ensure easy births for pregnant wives (2003:747).' My interpretation is that as men participate in *Sabarimala*, they have not only their wives and children in mind, but also their extended family. The fortunes of the members of the extended family, male and female, young and old, are borne to *Sabari-*

*mala* by family males. In this sense the pilgrimage can be seen as celebration of not only the male's role in the family but of family solidarity.

Of all of the rituals which emphasise family solidarity and continuity, perhaps the most important is that associated with pregnancy and the birth of the first child. In the next section, I relate the ways family is exercised in Govindan's daughter Meena's first pregnancy.

## Meena's first pregnancy

The rituals surrounding a woman's first pregnancy and childbirth involve an important reinforcement of family ties in the extended families of both the husband and wife. After their marriage, Govindan's daughter Meena and her husband Dinu moved into Dinu's parent's home in Neyyatinkara, a small town about an hour's drive south of Trivandrum. This practice of spending the first months after marriage in the home of the groom's family is quite common for families of all religions and castes. After about a year, Meena became pregnant. This initiated a series of ritual visits and moves by Meena between her and her husband's family. These practices evidently have a long history. Puthenkalam (1977) and Goughs' (1962a) descriptions of first pregnancy rituals resemble very closely those I observed after Meena became pregnant in 2001.

The first set of family reinforcing practices involves the announcement of the pregnancy. Because the early phase of pregnancy involves high risks, the husband and his family are not formally informed of the pregnancy until the fifth month. Dinu's family were formally informed by Meena's father Govindan, and they then responded by organising a formal visit to Govindan's home in Trivandrum. The visiting party consisted of Dinu's uncles and their wives, his brother-in-law (oldest sister's husband), and his sisters. The members of the visiting party dressed formally, the women in formal *sari* and the men in *munde*. They arrived by auto rickshaw bringing gifts consisting of five different foods to commemorate each of the first five months of pregnancy.[4] Meenakshi and the other women of the Govindan household had spent the morning preparing the meal that the two families would share, consisting of *sambars*, several different curries, *pappadam*, rice, and the popular Kerala dessert, *payasam*. At the feast, the men of Dinu's family were served first, followed by the women visitors, followed by the men of Meena's family. The women of Govindan's family did not eat until the guests had departed.

Meena and Dinu returned to his parent's home after the feast where they resided until Meena's seventh month of pregnancy. At that time, a visiting party from Govindan's family travelled to Neyyatinkara to pick up Meena and take her back to her family home. The visiting party arrived at Dinu's home with seven snacks to commemorate each month of pregnancy. Dinu's mother and other family women prepared and served another feast. Once again, visiting male kin were served first, followed by Dinu's family's males and then the women. After the meal, Meena approached Dinu's father and formally asked his permission to leave the household to return to her family home for the birth preparations. Her father-in-law prayed for her and gave his blessing. Govindan told me later that in his generation, after the wife returned to her parent's family, the husband was prohibited from seeing her until after the birth. These days this prohibition is disregarded. The husband can visit his wife as often as he wishes. Dinu visited Meena at least twice a month until she gave birth.

A special room was prepared in Govindan's home for the childbirth. For Meenakshi's generation, women usually gave birth at home. Today, almost all births take place in a hospital.[5] Meena gave birth to a daughter in a Trivandrum hospital and about a week later she and the baby returned to her bedroom in Govindan's home, where she remained for two weeks.[6] Three months after the birth, Meenakshi accompanied Meena and her baby daughter back to Dinu's family home. Soon after the return of Meena and her daughter, their new nuclear family moved from the Dinu's parental home into their own separate dwelling on a property adjacent to Dinu's parent's house.

This and other family rituals have importance at two levels. First, they illustrate how consumption, in the form of the movement and sharing of foods, are implicated in the reinforcement of family ties. Second and more importantly is that they constitute yet another set of practices invoking family loyalties and obligations. In Chapter 6 I will argue that the pull of the extended family is one of the important components of the process through which goods, ideas and new consumption practices move through work migrant families back into Kerala. Before developing the ways migration affects consumption, it is important to complete the exploration of dowry begun in Chapter 3. I return to Meena's dowry (introduced in Chapter 3) emphasising the ways in which dowry and consumption are mutually implicated.

## How dowry is implicated in household economy and consumption

Marriage and dowry involve significant exchanges of goods, money and property. Accumulating and distributing dowry involves an engagement of the extended family networks of both the bride and groom. In this section, the constituents of dowry are analysed, as well as how dowry is changing and how these changes are implicated in consumption.

An important point in understanding dowry exchange is that it is illegal. Governments have regarded dowry as primitive or premodern; therefore, dowry has been the target of legal reforms. Nonetheless, dowry is a flourishing institution in south India. The former leader of the powerful student union (SFI), now a lawyer, told me that when she was at university in the 1990s, SFI members vowed to eliminate what they considered the 'archaic' practice of dowry which was beginning to resemble a 'husband price'. The men vowed that they would never ask for nor accept dowry, the women that they would never let their families pay dowry. She said that eight years later, very few had kept their vows. I got the impression from her and many other people that they see dowry as an inevitable and integral part of marriage, though many upper middle class and elite families are reluctant to discuss dowry or to admit its importance in marriage.

Because dowry is illegal and regarded as premodern, dowry negotiations tend to be couched in the language of gift (a marriage gift is legal). Families approach specific demands for money and goods obliquely and euphemistically (see Bourdieu 1998 for a discussion of the use of euphemism in social exchanges). An approach from a potential groom's family in which a blatant demand is made for a specific amount of dowry could result in the bride's family breaking off the negotiations. Rija, the 24-year old Ezhava whose family was just beginning their search for her husband, told me: 'If the boy says "we want that land, give that house", we would not associate with that person. If he asks "give whatever you can give" then that is alright.' Rajan, a 22-year old male from an upper-income Nair family said 'we don't accept money like that (demanded). If after marrying, after a few years, if they give something, like that, money or a car or something, ok'. And Savita, a 20-year old from an upper-income Ezhava family who had spent most of her childhood in the Gulf, told me:

> My father has no intention of getting me married to someone who demands dowry. When he gets me married off he will definitely give something, but not the formal dowry thing. If you look at the society, I think the lower half would negotiate, but I don't think our family, if the boy said I want this, would respond. I think the girl's family takes it for granted that you have to give something when

you get married. So the question of the boy's family asking something, or a specified amount, I don't think we would respond.

Visvanathan (1989 and 2002) studied dowry practices among Christians and found the practices similar to those of Hindu Nair and Ezhava.

Examining dowry in 408 families surveyed, the differences in the components and sizes of Christians, Hindu Nair and Hindu Ezhava dowries were insignificant. Dowry remains an important cultural practice shared amongst all religions and castes in Trivandrum. The first time I talked with Abraham and Mea, a young Syrian Christian couple, about dowry, Abraham told me flatly that 'our line of boys don't ask for dowry'. He said that in the Syrian Christian families 'sometimes the boy's family gives dowry and in other cases they give whatever they want (again, he distinguishes between negotiated dowry and gift)'. However, in subsequent conversations, he mentioned numerous marriages in which large sums of money and goods had been exchanged in dowry. In the end, Abraham characterised dowry as a social reality that 'you can't fight'.

Abraham: When there is a wedding the first thing people ask about is 'how much?'
Harold: I wonder what is driving that?
Abraham: The driving dynamic is that he (the groom) gets about 10 *lakh* and puts it in the bank and he gets about a *lakh* and a half in interest. Then that is like an additional boost to his own income. It is a little cushioning. For a little more cushioning it is my opinion that they are doing something that should not have happened. It is very hard to say, from our side, from my sister or my cousins, we have instances where we have given also. But a reasonable amount. It is a social setup, you can't fight it, so we let it be.

## Meena's dowry

The story of Govindan's oldest daughter Meena's search for a husband illustrates many of the problems and subtleties of dowry negotiations for a lower middle class family. As is the case in virtually all Kerala marriages, the formal responsibility for finding the partner and for negotiating dowry lies with the men of the family. Govindan had the formal responsibility for finding husbands for both of his daughters, Meena and Lakshmi. He was assisted by his son Anil. However, the senior females in the family (his sister Rajama and the wives of Meenakshi's brothers) were consulted at each stage of the dowry negotiations and played an important role in either accepting or vetoing a negotiated dowry.

The search for a partner for Meena began shortly after she turned 20. Only men of the Nair caste were considered. Anil told me that the most important criteria for the groom were that he be well-educated, employed, a few years older than Meena and that they be astrologically compatible. When dowry negotiations had come far enough, the astrological charts of the candidates for Meena and the potential groom were prepared by a priest or astrologer. A judgement that the couple is astrologically incompatible invariably puts an end to the process. The chart also indicates auspicious and inauspicious periods for marriage. In Meena's case, after the first few partner negotiations had failed, she was advised by the astrologer not to marry between the ages of 28 and 32.5. Since the family had not found a husband for Meena before she turned 28, the search was suspended until she was almost 33.

During the first few years of the search for a husband for Meena, suggestions for candidates came mainly from one of the members of the extended family. Background checks on a few suitable candidates were done, and in a few cases discussions about dowry were initiated. However, no suitable candidate emerged; therefore, Govindan and Anil decided to contact a 'marriage broker'. Marriage brokers bring families of brides and grooms together and supervise dowry negotiations. They demand a fee, usually about 5 per cent of the value of the negotiated dowry. After a few unsuccessful attempts, Govindan gave up on the broker. According to Anil, the broker tended to exaggerate the qualities of the potential groom. Also, using an intermediary made it harder for Govindan and Anil to assess the character of the groom and his family. In Meena's case, after several brokers failed to find suitable candidates, Govindan fell back on the family network as a source of potential grooms.

Over the six or seven years of searching for a groom, Govindan entered dowry negotiations with 15 families. Govindan and Anil said that most of the negotiations fell through when they realised that the dowry demands by the groom's family would be beyond what Govindan's family were willing, or able, to pay. In the final successful negotiation, Anil told me that the key to success was a less aggressive attitude towards dowry on the part of the groom's family. The groom's father indicated at the first contact that he would be satisfied with a moderate amount of gold and property. He did not mention cash. Govindan was relieved because all of his land was mortgaged; therefore, directly transferring property, or securing a loan for purchasing gold or providing cash was out of the question. Further, the groom's father agreed to postpone the land transfer for a year, giving Govindan time to pay off the mortgage. Not all of the senior members of the groom's family were happy with this condition. One of the groom's uncles insisted that the land be transferred at once. According to Anil, this demand came close to sabotaging the marriage arrangement. Word got back to Govindan about this controversy in the groom's family and he immediately cancelled the agreement out of pride.

Meena and her family were obviously distressed about the prospect of another failure and further delays. Meena was by then over 33 years old and had no new prospects in sight. As a result, at a subsequent crisis meeting in the family, Meena's maternal uncles decided to step in. They offered a piece of property adjacent to the property originally promised. Govindan, who would eventually have to pay back his brothers-in-law, went along with this, as did the groom's family. An engagement date was fixed and the wedding took place six months after the engagement. In addition to the negotiated property and jewellery, 50,000 rupees was promised as 'pocket money'. An important point is that the groom's family had not demanded cash but they expected 'pocket money'. 'Pocket money' is a euphemism in two senses, first because its exchange is couched in the terminology of gift and second because it gives the impression of being a small amount that can be slipped into the groom's pocket rather than the tens of thousands of rupees that is usually exchanged.

In addition to dowry, Govindan's family agreed to pay the expenses connected with the engagement and wedding ceremonies. He also provided expensive saris for the groom's mother, sister and sisters-in-laws.

Once Meena's situation was resolved, the search for a husband for her sister Lakshmi was intensified. As is typical in the cases of younger sisters, her prospects were weakened by the economic strains on the family brought on by Meena's marriage and dowry. After several unsuccessful negotiations with the families of potential grooms, the family was thrilled in April 2002 to find a candidate who announced that he was not interested in dowry. A meeting was set up between the senior men of both families. Govindan told me with a touch of irony that the first question the groom's father posed was, 'What will you give with her?' Govindan responded with an offer of gold and a promise to stage an elaborate wedding 'consistent with the status of the groom', who was the owner of a small shop. This was agreed to by both families and a meeting was set up between Lakshmi and the potential groom. They were both amenable to the marriage. Their astrological compatibility was checked and found to be acceptable. The traditional meeting of the two families to celebrate the agreement followed soon thereafter at Lakshmi's home, where 30 members of the groom's family were served a lavish meal.

Unfortunately, this chapter of Lakshmi's story came to a sad end soon after the feast sealing the engagement. Anil discovered that the groom had misrepresented his job situation. He was not the shop owner, as he had claimed, but was actually a salaried employee at the shop. Because of the deception, the marriage plans were cancelled. Lakshmi and her family were back to square one. She still faced a continuing existence as family ward and the looming identity of spinster, as well as a continuation of the exhausting rounds of search and negotiation. Fortunately, two years later a match was made which did work out and Lakshmi was married in 2004.

The story of the husband search and dowry negotiations by the family of Meena and Lakshmi is similar to stories I heard from many other middle-income Hindu families. Exorbitant demands and extended negotiations subject the family to economic and social pressures which extend for years. In Meena's case their difficulties in finding a groom and managing dowry demands was a source of frustration and embarrassment for Meena and her family.

Among the few Muslim families interviewed, I also found the same tendency to initially deny that dowry was involved in either their own or their children's marriages. In subsequent conversations people would usually reveal, one after another, things that accompanied the bride. One example is a Muslim family I interviewed in which the husband and wife are both university teachers approaching retirement. They both said initially that they 'did not believe in dowry'. Later, as they described their daughter's wedding, they gradually revealed what turned out to be a sizeable dowry. It included a house as well as many of the furnishings, including a refrigerator, television, and an expensive dining room table. Their daughter asked if the parents could provide 'an a/c room' for their new home, so the parent's bought and installed an air conditioner. When I asked for an estimate of the total expense, they estimated that the house and furnishings cost them 2.5 million rupees. As an afterthought, they mentioned that the groom had also asked for an 'a/c car'. They bought a used car for 40,000 rupees. They also included a 30,000 rupee motorcycle, intended for use by their daughter. The wedding ceremony cost them 40,000 rupees. Thus the total cost to this family, who initially stated that they did not believe in dowry, was in excess of 2.6 million rupees. When I asked them how they managed this expense on their teachers' salaries, the wife responded, 'We suffered for many years'. They told me that they had worked hard, kept their expenses low and lived 'Spartan lives' in anticipation of the costs of their daughter's dowry and wedding.

These examples reveal that dowry is both socially important and economically significant. In the next section, the constituents of dowry are examined, how they are changing and what the changes mean for consumption.

**The constituents of dowry**

From the mid-20th century, land and cash joined jewellery as regular components of dowry. The amounts of land and jewellery in dowry grew after the 1950s, probably attributable to land reforms and a rapid increase in repatriated Gulf income. In Table 5.1, dowry reported by households who participated in the survey is aggregated in ten year intervals, from 1957 to 1996 with a final interval of five years from the period 1997–2002. In order to obtain more information on the makeup of dowry, questions about dowry were included on the survey questionnaire. One question asked for the amount and composition of dowry transferred in the marriage of the male and female heads of household. A second asked whether a close family member (brother, sister, cousin, nephew or niece) had been married in the last five years, and if so, what the size and composition of the dowry had been. The specific family marriage could not be identified in this latter question. Comparing the responses to the two questions (heads of household married in the past five years, and some other family marriage in the past five years), the cash reportedly transferred in the latter cases was three times that reportedly transferred in the former cases. The amounts of gold reportedly transferred were also higher in the latter cases. This supports my impression that people play down the amounts of cash in dowry.

In addition to cash, gold and land, a house was transferred in dowry in about one-fourth of marriages in each of these time periods. As Table 5.1 reveals, the amount of land transferred has varied considerably over the

*Table 5.1* Average cash, gold and land transferred in dowry, based on a survey of 408 households in four Trivandrum neighbourhoods

| Dowry in | 1957–66 N = 46 | 1967–76 N = 91 | 1977–86 N = 124 | 1987–96 N = 72 | 1997–2002 N = 26 |
|---|---|---|---|---|---|
| Cash in rupees | 39,167 | 54,300 | 33,000 | 38,800 | 33,340 |
| Gold in sovereigns | 25 | 29 | 35 | 39 | 46 |
| Land in cents | Lack of data | 31 | 28 | 132 | 32 |

One sovereign (sorin) is eight grams of gold, worth about 4600 rupees in 2002.
One cent of land is 40.47 sq meters

years. The table reveals a peak in land transfer in the period from 1987–96, during which the price of land increased sharply in anticipation of the 'opening' of India to foreign investment in 1991. From the mid-1990s land prices stabilised and then fell somewhat; likewise the amount of land transferred in dowry stabilised and then decreased somewhat over the same period.

Cash dowry increased rapidly in the late 1960s and early 1970s in Trivandrum and elsewhere in south Kerala (Lindberg 2001). This increase is likely related to the first wave of repatriated income from Gulf work migration. After the 1970s, cash dowry has been fairly stable, yet Table 5.1 shows a steady increase in the amounts of gold in dowries. Gold is popular because it has not only a cosmetic and visual usage when worn as jewellery, but it is liquid, in that it can be easily converted into cash. This can be advantageous for the groom's family, who can then use the dowry to supplement the household economy or as a part of dowry for family women.[7]

An important change in dowry over the last couple of decades is the inclusion of household furnishings, cars and household appliances such as televisions, refrigerators, washing machines, microwave ovens. One reason for the popularity of these goods is that they can be claimed as gifts and thus are legal. Another reason is their visibility. Syrian Christian Abraham told me that his cousin had insisted on having a sports utility vehicle (SUV) as part of her dowry, even though she and her husband would be moving to the United States after the wedding, and would therefore have to sell it shortly after the marriage. They wanted the SUV because it was 'flashy', in Abraham's words, and because it could readily be sold for cash.

Cars share with gold the characteristics of visibility and liquidity – the relative ease with which they can be sold and converted to cash. There is a vibrant market in Kerala for used cars, with easy access to information, dealers and individual sellers. There is a much less developed market for used household appliances and no regularised routines for buying and selling them. They depreciate considerably in value after they are used and are more difficult to be sold or pawned. Thus household appliances are not as liquid as gold and cars, making them somewhat less attractive for families of grooms. However, from the perspective of the bride's family, this lack of liquidity is seen as positive. The household appliances, which will be available for use by the new wife in her household chores, will make life easier for her. In other words, she will be able to benefit from her dowry. Including appliances in dowry thus reinvests dowry with one of its original purposes, namely to provide the bride with things that can be of use to her in her new home.

## The embedded cost of education

In the discussion of marriage partner selection in Chapter 3, it was pointed out that a university education makes a woman more attractive as a potential mate.[8] Highly educated men are also more attractive as mates and can therefore make higher dowry demands. Thus, ironically, one of the reasons why families invest in their children's education in anticipation of marriage partner searches and dowry. This investment is costly. It begins with kindergarten fees, which range from 6000 to 20,000 rupees yearly per child in Trivandrum. Private evening tutorials for children are quite common for middle class children and cost around 60 rupees per hour (on average 8000 rupees yearly per child). In public schools, yearly educational fees from first grade to post-degree (high school) are around 4000 rupees yearly. Recently, many people are electing to put their children in expensive English-language schools (up to 100,000 rupees yearly per student in 2002), partly because English skills increase competitiveness for university admission. University tuitions range from 10,000 rupees per semester in public universities to 50,000 rupees in some private universities.[9] There are also unofficial costs. A separate donation (a euphemism for bribe) is expected in order to assure acceptance of one's children at all private schools from kindergarten and upwards. At the university level the expected donation can be as much as 100,000 rupees per student.

The total educational costs for boys and girls who pursue a university degree add up to a considerable sum. The socio-economics of dowry are such that the parents of grooms can expect to recover their education costs through dowry, while the parents of brides cannot.

## Marshalling dowry

Accumulating dowry is a lifetime project, involving saving, borrowing, pawning jewellery and selling off pieces of family property. Parents with more daughters than sons, said things like, 'We suffered for many years' when describing the hardships of preparing for dowry and education costs. These hardships are greater for lower and middle income families. Table 5.2 shows average amounts of cash, gold and land transferred among low, middle and high income families who participated in the survey.

Converting gold to its equivalent cash value (one sovereign equivalent to 4600 rupees in 2002), the sum of gold and cash transferred for lower, middle and higher income families 179,000 rupees, 257,100 rupees and 222,801 rupees, respectively. For lower income families, this cash value alone is equivalent to almost four years of income and for middle income

*Table 5.2* Average cash, gold and property components of dowry transferred according to the monthly income of households, based on a survey of 408 households in four Trivandrum neighbourhoods

|  | Lower middle income: 1000–7000 rupees N = 132 | Middle income: 7001–15,000 rupees N = 118 | Upper middle and wealthy: >15,000 rupees N = 128 |
| --- | --- | --- | --- |
| Cash in rupees | 41,000 | 56,100 | 34,201 |
| Gold in sovereigns | 28 | 36 | 41 |
| Land in cents | 63 | 109 | 582 |

families cash value is about two years of income. Comparing the cash and gold transferred, the amounts in the dowries of middle income families are actually higher than those in higher income families. Higher income families transfer more land, but these are usually in the form of holdings (capital). In general, one can conclude that the economic strains of dowry are greatest for low and middle income families.

From the first time I was exposed to the economics of dowry, I puzzled over how families managed to come up with the large amounts of cash and goods. I expressed my puzzlement to an architect and home builder with long exposure to the economics of Trivandrum families. He said, 'You have been here just eight months and you are amazed by it; I have been here all my life and I am still amazed. It is something which beats the economist.' He speculated that part of the explanation may lie in the changes in land ownership which can be traced back to land reforms of the 1960s and 70s.

> I don't know how it works, but all these persons probably come from a family which over the years has not really destroyed its own access to wealth so that this guy's income is supplemented by income probably from his land or his house, or he has got two houses, or his wife has land holdings which has come through dowry. All this is because there has been a very steep appreciation in land cost which permits them to sell splices of this land.

His hypothesis about the ways that families strategically use land holdings may be a key to understanding how families manage to come up with such huge dowries. In Trivandrum, abandoned agricultural land which had earlier been used to grow paddy was very cheap in the 1960s. Members of Govindan Nair's extended family, for example, bought a lot of paddy fields which they have sold as needed in subsequent years.

Several other families told me that they sold their own homes to provide dowries for their daughters and then later repurchased a new home with dowries paid to their sons. A few families sold their homes in Trivandrum and purchased less expensive homes in nearby villages. Virtually every family I interviewed had strategically used buying or selling of property, owned either by themselves or someone else in their extended family, in putting together the dowries of family women.

## Conclusion

Simpson writes that family ought to be reinserted into consumption theory in Europe because it provides 'the web of meaning upon which the most fundamental of human relationships appear to be hung (1998: 11)'. The evidence from Kerala affirms the relevance of this observation there. Despite intergenerational and intergender conflicts about and within families, people believe in the extended family, exercise family relations and use those relations in acquiring goods.[10] Family has fundamental social importance, exercised in events, rituals and especially in marriage. The extended family network is actively exercised through the celebration of important life cycle events in the lives of its members, through regular visits, both informal and formal, and through parental and sibling relationships that involve mutual obligations and mutual care. Family remains an ideal and a template in the mediation of all kinds of actions, including consumption. In the realms of family economics, filial obligation, sibling relationships, marriage, dowry and the passing on of family traditions in child-rearing and cooking, the ideal of family remains strong, as does the social network it represents.

Dowry is a family exercise and today's dowry *is* a form for consumption in that it involves the acquisition of gold, property, and increasingly cars and household appliances. The consuming unit in dowry is the extended family. Goods and money flow through family while commodities such as the household appliances drop out of the flow and into use in one of its constituent households. Thus the acquisition of goods is a social process without spatial limitations. As will be explored in the next chapter, family networks act as conduits for commodities from the Gulf and, equally importantly, conduits for new ideas about consumption and the role of consumption in household practices. Thus consumption in Kerala is deeply implicated in the extended family, dowry and work migration.

# 6
# Work Migration

There are approximately 1.5 million Kerala residents who maintain an address in Kerala but whose principle place of work is outside India (Zachariah and Kannon 2002). Kerala work migration has its roots in the early 20<sup>th</sup> century. Families marshalled their resources and sent one male family member in search of a job. For Malayalee, the early destinations were Malaysia and Singapore, but from the 1960s huge numbers of migrants found work in the Gulf. The population of Malayalee migrant workers grew rapidly first in Kuwait and later in Saudi Arabia, the United Arab Emirates (UAE), Bahrain and Qatar. More recently migrant workers have increasingly looked beyond the Gulf, to Australia, Europe and North America. According to Zachariah, Mathew and Rajan (2002a), work migration to the United States tripled during the period from 1999–2004.

This chapter will explore the ways in which consumption is implicated in Kerala work migration. Work migration extends the family network beyond national borders and facilitates the transnational movement of money, goods and ideas about consumption. Appadurai (1996) used the term 'ethnoscape' to capture the transboundedness of this new global reality; others, such as Olwig (2003) and Kearny (1995) use 'transnational field'. An element in both of these conceptualisations is a fluid sense of place and identity, as well a transnational movement of both ideas and things. Appadurai introduces the notion of 'workscapes' and 'familyscapes' that span two widely separated locations, both geographically and culturally. These 'scapes' are important conduits for change in household consumption practices in Kerala.

With a few exceptions, notably Gulati's work on women of migrant families (see 2003), most scholarly works on work migrants have focused on the socio-economic sides of migration and on the repatriation of income. This is understandable because the volume of repatriated income is huge. Zachariah and Kannon (2002) estimate that Non-resident Indian (NRI) foreign remittances to Kerala in 1998 totalled 138 million rupees (USD 69 million), of which 126 million rupees were remitted from the Gulf countries. They claim that for 1998 this constituted 20–30 per cent of Kerala's gross domestic product.[1] Furthermore, they found that 'More than a million families depend on migrants earnings for subsistence, children's education and other economic requirements (2002:2).' Kerala is one of several places in Asia in which the migration involves sizable numbers of people and their families. Other countries which have experienced large movements of people seeking work outside their borders are the Philippines, Bangladesh, Sri Lanka and Indonesia. Johnson (1998) studied the lives of Filipinos who were work migrants from the city of Jolo to the Gulf countries. He found that 39 per cent of families in Jolo had a family member working in the Gulf, similar to the situation in Trivandrum. Work migration is an important transnational phenomenon in Asian countries, and given the global restructuring and fracturing of commercial enterprises, it is likely to continue to grow in importance in many places around the globe.

South Kerala work migrants are mainly Hindus and Christians. Therefore, south Kerala migration to the Muslim-dominated countries of the Gulf differs from migration from northern Kerala, Bangladesh and Indonesia, where the majority of migrants are Muslim. Muslim work migrants to the Gulf move between Muslim-dominated societies which share religious and cultural practices. However, social differences between home and place of work are much greater for the south Kerala Hindu and Christian migrants to the Gulf. Southern Kerala migrants tend to mix with other Malayalee and Indians in their countries of work. Unmarried migrants usually live in gender segregated apartment complexes and married migrants in housing communities built especially for migrant workers.

Another unique aspect of south Kerala migration is that the majority of migrant workers are men. For other parts of India and Asia the proportion of women migrants is higher. According to Zachariah, Mathew and Rajan (2002b), at the end of the 1990s, only about 10 per cent of Kerala work migrants were women. Another difference is in the nature of the work abroad. While work migrants from other Asian countries do mainly work as domestic servants and other forms for unskilled labour, most of migrants from Trivandrum work in jobs that require higher education and technical skills, such as jobs in the health care and oil industries. Partly as a consequence of the prospect of higher paid employment, migration of whole families from Kerala was fairly common in the migrations of the 1960s and 1970s. There has been a decline in whole-family migration in recent years, partly because availability of jobs in general declined after the first Gulf War in 1991, but also because the work contracts have on average been shortened (Zachariah and Rajan 2004). In recent years, it is more common for young males to migrate, but marry women from their communities in South Kerala. After marriage, the men only return once or twice a year to be with their wives and children. In the initial years after marriage, it is common for the wife and children to reside with either her or the husband's parents. As the husband begins to accumulate capital, the family begins construction on its own house in Kerala. The wife and children then occupy the house when it is completed. When the husband retires, it is common he and his family to settle into the Kerala house.

One neighbouring family in Kumarapuram, headed by Binu and Cavita (not to be confused with Chavita, the Hindu Nair school teacher whose typical day was described in Chapter 4) recently retired to Kerala after spending much of their working lives in Kuwait. Their story and that of their now adult children who were raised in Kuwait brings to life some important ways that migration engages family and contributes to changing ideas about consumption.

## Cavita's migration story

Cavita and Binu are Hindu Ezhava, both in their seventies. They migrated separately to Kuwait around 1960. They moved back to Trivandrum in 1990 to a home they had constructed over a 20-year period commencing in 1970. Their two sons, in their twenties when their parents returned to Trivandrum, had until that time only experienced Kerala as a place where they spent brief vacations.

Cavita had been raised on the family farm in south-central Kerala, along with 'sisters and brothers and everybody'. Her family encountered economic difficulties after her father took over management of the farm. What had been a productive farm with many tenants and considerable income from cash crops began to decline when Cavita was a teenager. A decline in the prices of cashew and rice paddy in the 1950s contributed to the decline in the farm's fortunes. Cavita said that her brothers looked to new work opportunities in the Gulf 'to find a way to help him (their father).'

Cavita's older brother was in the first group of Malayalee to seek and find work in Kuwait. He fulfilled his promise to his father by regularly sending back money, but he had two daughters who would need dowries, so much of his Gulf income was set aside for their marriages. Since Cavita was a woman, she did not give much consideration to work abroad. This changed when she had completed her high school education and she and her family were paid a visit by one of her former teachers. The teacher had moved to Bombay (Mumbai today) to live with her father, who made a living assisting Malayalees to find work in Kuwait. Bombay was the principle jumping off point for Gulf workers. Cavita's former teacher told her about the growing demand for nurses in Kuwait. She encouraged Cavita to come to Bombay and enrol in a nursing school, where she could use the teacher's father's contacts to secure a nursing job abroad. Cavita followed her advice. She enrolled in a nursing school, studied English in the evenings (a requirement for obtaining work in Kuwait) and at the end of her schooling had a job interview with a Kuwait employer. She did not immediately receive a job offer, so she took her nursing certificate and returned home. However, nine months later she was contacted and offered a job in a Kuwaiti hospital. In May 1960 at age 30 she left for Kuwait, the first woman from her village to do so.

The conditions she met on arrival were tough. She and several other nurses were given a tent to share. The tent was extremely hot and exposed to desert winds. When she returned home after work, she often found the tent collapsed and covered in sand. In the meantime, she met a young man named Binu, from a village near hers in Kerala. Binu was employed by a European-based oil company. Cavita and Binu became acquainted and after a few years decided to approach their parents about marriage. Their parents initiated the process of checking horoscopes and negotiating dowry. When both families had approved the marriage, Binu and Cavita returned to Kerala for the marriage ceremony, which Cavita said was 'the rule' for couples who met overseas. After the marriage, they set up a household in a small apartment in Kuwait. By then conditions for migrants had improved considerably. Cavita and Binu were given their own flat. Cavita described it this way:

> All modern things were there. We were cooking on our own. The company supplied everything, gas, water. We didn't have to pay for anything. It was very expensive, the water, but we didn't have to

pay. Everything was free. Only the things we have to cook (we paid for).

She set up housekeeping using her modern appliances. Cavita said that she and other work migrants worked hard and lived frugally.

> We suffered a lot. We used to work from 7 to 7. My husband had night shifts…We went to (Kuwait) to make money. Money and all of these things which have been bought from there (pointing to her Italian cooking range, refrigerator, stereo, air conditioner and so on).

Cavita and Binu worked long hours and saved money for 'all of these things', but there were two other destinations for their income: members of their extended families in Kerala and the entrepreneur they had hired to supervise the building of their retirement house in Trivandrum. They sent money regularly to both sets of parents and senior members of both families. They provided cash and household goods for the dowries of cousins and nieces, as well as for wedding gifts for other members of the extended family. They were frequently approached for loans by both family members and their in-laws. On each of their trips back to Kerala, they brought goods intended for members of the extended family. Chavita's son Gopal, told me that he and his two brothers, both of whom also work (or have worked) in the Gulf had carefully set aside gold and cash for the anticipated demands for marriage gifts and other family commitments they knew would be asked of them by family relatives, both close and distant. In the Gulf families I got to know I observed that if there was a financial problem in the family or if there was a female approaching marriageable age, the family tended to turn to the Gulf worker. Shoba's husband Thomas told me that his returns to Kerala were characterised by a procession of family members (especially his sisters, both married) who approached him for economic support. There is an expectation that work migrants bring gifts for virtually everyone in the extended family on their frequent returns. Thomas told me he was expected to bring cell phones, dresses, perfumes, tape recorders and electric appliances, even if they could be obtained at an equivalent or better price in Kerala.

Cavita called her house construction project in Trivandrum a 'money sink'. She said that things inevitably cost more than they expected, because they were not there to supervise the progress of the construction. There were many construction delays. They realised that many of these delays were due to the fact that bribes were expected but not forthcoming. They finally understood that in order to expedite the building, they would be obliged to pay bribes to the contractor, city planning officials and others to ensure that construction progressed according to plan.

To meet the demands from their Kerala families and from the costs of constructing their home in Trivandrum, Binu and Cavita worked long hours six days a week. These long hours, and their new household appliances, together contributed to important changes in household practices regarding cleaning and cooking. They bought a used car to commute and to shop. Cavita's washing machine permitted her to wash clothes after work in the evenings and on Sundays, her day off. Cavita continued to prepare typical Kerala meals, bringing ingredients back with them on their frequent visits to Kerala. However, in a departure from Kerala practices, she began cooking dishes in bulk and storing portions in the refrigerator for later reheating. She prepared a whole week's worth of meals every Sunday, and then reheated them for weekday meals. These new consumption practices, developed under conditions Cavita encountered in Kuwait, were eventually transferred back to their homes and consumption practices in Kerala. I return to this important point below. First, I address stereotypes about work migrants and how these have affected the ways that migrant families consume.

## People from nowhere

Cavita's story illustrates some of the ways that 'easy money' from the Gulf had its costs in terms of hard work and demands from the extended family. In their descriptions of Gulf workers, non-migrants tended to emphasise the 'easy money' migrant workers earn and how they use it to draw attention to themselves. Gulf families are said to use their money conspicuously and unproductively. Zachariah, Mathews and Rajan (2002b:91) write that Gulf migrants used their money on consumption rather than on 'nation building activities' such as investments in 'technology infrastructure'. This 'misuse' of wealth carries with it an implication that migrant workers are self-centred and unpatriotic. Johnson describes a similar view of Gulf migrants in the Philippines, both by the people he interviewed and in scholarly articles. He reviews articles on migration (Arcinas 1986 and Jackson 1990) and writes that the authors accuse migrants of not using their overseas remittances 'on local investment in productive enterprises, creating a new system of dependency based around overseas earnings (1998: 230).'

In Kerala, many families without a Gulf connection attribute the inflation of dowry to the easy money and goods brought back to Kerala by migrant workers. The survey of four Trivandrum neighbourhoods shows that there were no significant differences between the cash amounts in dowries of Gulf and non-Gulf families; however, Gulf families are more likely to include household appliances and cars in dowry, and this has undoubtedly influenced the dowry market.

Kerala historian K. Saradamoni (interview) argues, as do others, that the image of the easy life of Gulf migrants is unfair and incorrect. There are many hardships associated with migration which tend to be ignored or forgotten. Saradamoni points to the significant amounts of money families had to marshal to finance a Gulf venture by one of its members. Zachariah, Mathews and Rajan estimate that an average NRI in the 1990s needed 44,000 rupees just to get to their country of destination (2002a:18). Money went to tickets, visa fees, and agents, and to bribes and other hidden fees. There are many stories of hopeful migrants who got as far as Mumbai, the jumping off point for overseas jobs, but failed to secure a job and returned to Kerala empty handed. There are also stories of those who made it to the Gulf, only to have their work contract shortened or abruptly terminated. Zachariah, Mathews and Kanon found that 20 per cent of migration attempts from Kerala were 'misadventures', resulting in 'loss of wealth, wastage of energy and loss of health' (2002a:6).

Our friend Anil was one of those who got no further than Mumbai. Like Cavita and many others, he went to Mumbai seeking help to find work in the Gulf. He told of having to scrape out a living there while he waited for a job offer. He gave up when his money ran out, forced to return to Trivandrum with no job and with his family's seed money gone. Anil's entire extended family was involved in financing his venture, an involvement which Gulati (2003) found to be typical of the early period of work migration. Anil's family had borrowed money and mortgaged land to support his effort. Anil said that his family and friends treated him as a failure on his return. Rija's uncle Akbar was another who felt he (Akbar) had failed his family. Akbar had secured a job in United Arab Republic, but was forced to return home long before he had intended when his contract was abruptly terminated. He suffered through three years relying on the good will of his joint family while trying to find another job outside India. Later he gave up hope and searched in the much less lucrative job market in Kerala. He still lives in the joint family household, working in poorly paid part-time jobs. The failures of Anil and Akbar are not uncommon, especially of the period after the mid-1990s. It has become harder to find work in the Gulf; conditions are less appealing and contracts are often for short durations.

Cavita's family moved back to Kerala just before the event that was to dramatically change conditions for migrants in Kuwait and other parts of the Gulf, the Iraqi invasion of Kuwait in 1991.[2] Workers' funds were stranded in Kuwaiti banks. When order was restored after the war, they were able to get their money reimbursed, but they were not compensated for their loss of property or belongings. A few got their jobs back, but many jobs were eliminated as part of a new Kuwaiti strategy to make the economy less dependent on foreign labour.[3] In general, the lucrative work conditions for foreign workers in Kuwait and other Gulf countries declined after 1991. Malayalee workers are looking to other parts of the world, such as Great Britain, the United States, Canada and Australia for employment. Travel to these countries is more expensive and obtaining work visas more difficult. It is increasingly the wealthier families who can put together the capital to send one of their members abroad. In the mid-20th century, the migration of mainly middle and lower income families had the effect of reducing the gap between rich and poor; today it tends to reinforce the gap.

The stress of family separation is another hardship associated with work migration. Most Gulf families are in a state of constant separation, with the husband abroad and the wife and children at home. Most migrant husbands return only once or twice a year to visit their Trivandrum families. The full responsibility for the children and the Kerala household falls on the shoulders of the wife. Shoba is a Latin Christian whose husband Thomas works in the United Arab Emirates. He returns to Trivandrum twice yearly to be with his family for a few weeks each time. Shoba lives in a small two-room house with their two children, five and eight years old. She expressed how tough she found this situation to be. For one thing, she had earlier worked in Saudi Arabia, where she had shared a home with Thomas. Her contract as a nurse at a Saudi hospital was terminated in 1998 as a result of the change in Saudi policy which replaced foreigners with Saudis in health sector jobs. She was not able to find work in Trivandrum. The loss of her Gulf income weakened the family economy and resulted in a stop in the construction of their house in Trivandrum (it had been under construction for five years). They were forced to rent a small house for her and the boys. In 2004 she was still desperately searching for a job that will allow her and the children to rejoin her husband in Saudi Arabia. Chavita's sister Parvati (Hindu Nair) is another 'Gulf widow' (a term coined by Gulati 2003). Parvati told me that her husband had only returned to be with her family six times in the 12 years he had worked in the Gulf. When I asked her why they put up with this she answered, 'For the progress of the family. There is a daughter to get married, for the education of the children, he (her husband) constructed our house, and also to take care of his parents.'

Cavita's son Gopal was forced to return to Trivandrum when his job contract in Kuwait was not renewed. He has spent the last few years in Kerala, helping on the family farm (work which he said he disliked intensely) and 'surfing the net' for jobs abroad. He complained that the family's economic situation was 'not that good'. He blamed his parents for using too much of their Gulf wealth on their house and its furnishing, which had drained their savings. The family has been living off interest income from their savings and, after Gopal's marriage in 1995, on the added income of Gopal's wife Kahina's school teacher salary. Gopal was one of many children of early migrants I talked with who had been raised outside Kerala and who had never really felt 'at home' after their return to Trivandrum. He referred to himself as one of the 'people from nowhere':

> Because they (his parents) were so busy, we were not brought up in the Kerala tradition. I feel my daughter should have a more traditional Kerala upbringing. We don't know much about the traditional dances. Or about our own religion. She should know more, (about things like) the classical arts and she should know more know about her roots. Because we were brought up in a more Arab way, we don't know much about the traditional Kerala life. It is not only me (who feels this way). A good amount of our parents have gone to the Gulf. It is a sizable population who has been there and have come back. To integrate with the society we find it difficult. Because we don't know much of the traditions. And over there we lived in clusters and didn't really mix much with the Arab society. We don't have much of the Arab culture or the Kerala culture. We are people from nowhere.

A psychologist who works in family counselling told me that many of his clients were former Gulf workers, or children of Gulf workers. Many of them expressed this feeling of having a 'nowhere' identity. Because of the sense of dislocation, exacerbated by the persistent stereotypes discussed above – non-productive, consumerist, lazy – many migrants have undoubtedly used consumption as a vehicle to get from 'nowhere' to 'somewhere' in the Kerala social landscape. Social performance is a motive for consumption in migrant families, but my impression is that it is secondary to the importance of servicing obligations to members of their extended families. In the next section, the results of the survey questionnaire will be used to support this point.

## Consumption in 'Gulf' families

Responses to the survey questionnaire give a basis for comparing consumption between Gulf families and other families. In order to identify Gulf families, one of the survey questions asked the participants to indicate whether they had a close family member working 'outside India'. In the figures below, these families are referred to as 'NRI' (Non-resident Indian), the official designation for a work migrant. To ensure that participants in the interviews interpreted 'close family' correctly, we asked them to specify which of the following described the family relationship of the NRI to the female head of household: spouse; parent; child; sibling; a marriage partner of a child or sibling; uncle; nephew/niece; or a first cousin. Forty-one per cent of households surveyed in the four middle class neighbourhoods reported that they had a close member working outside India. These results accord well with Zachariah and Rajan's (2004) finding that 38 per cent of Trivandrum middle class households had someone in the family working outside Kerala.

One of the striking results of the survey is that Gulf families in these middle class neighbourhoods have on average double the income of their non-Gulf neighbours. Thus Gulf families are decidedly wealthier than their non-Gulf neighbours. Examining the size of the house and the appliances in it, a comparison reveals Gulf families have bigger houses and more household appliances. They are also more likely to have a car. Figure 6.1 compares ownership of selected household appliances and cars in NRI and non-NRI families.

Since incomes of NRI families are on average higher, differences of ownership could simply be attributed to differences in income. In order to test this, ownership of these items was compared across similar income categories ranging from low to high income. Four representative comparisons are illustrated in Figure 6.2.

These comparisons show that for households in the same income category, there is a consistent pattern of higher ownership of household appliances (and cars) among NRI families. If income differences are not responsible for these differences in consumption between Gulf and non-Gulf households, what is? I suggest that there are two related explanations, one having to do with the pull of the extended family and the other with the work migrant's sense of dual residence: one residence inside Kerala and the other outside India.

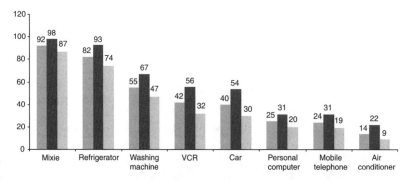

*Figure 6.1* Ownership of selected appliances comparing families with a family member working outside India (overseas) versus families with no family member working outside India (no overs), based on a survey of 408 households in four Trivandrum neighbourhoods

## The pull of family

We have seen how the extended family networks of migrants span large geographical distances. For Trivandrum work migrants, distance does not affect the strength of the extended family. The flow of goods and money from abroad through the Kerala extended family is used to repay parents and to contribute to the building of a home, but also for marriages, dowries, and the resolution of economic crises and so on for aunts, uncles, cousins, nieces and nephews. This is not to deny the there are disruptive aspects of migration. Husbands live lives apart from their nuclear families. This physical separation is one of the issues that lead early researchers on Indian (see Epstein 1962) and African work migration (see Murray 1981 and Brown 1983) to conclude that migration was responsible for weakening family or breaking it down altogether as a social force. Measured in terms of the emotional and practical problems that stem from physical separation or in terms of conflicts arising from the meeting of new ideas from lives lived abroad and traditional ways of doing things in Kerala, one could assert that families are weakened by work migration. Nonetheless, my view is that while these kinds of conflicts exist, they have not diminished the strength of family. In most Gulf families, workers have strong feelings of loyalty, as well as a sense of obligation and reciprocation to members of their extended families.[4] These social pulls are responsible for drawing goods and money through extended family

100 *Consumption and Transformation of Everyday Life*

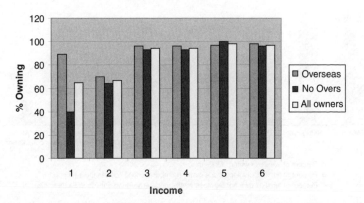

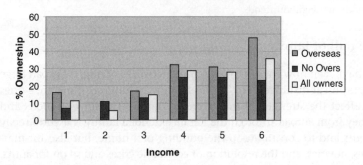

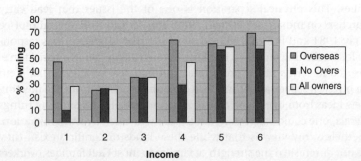

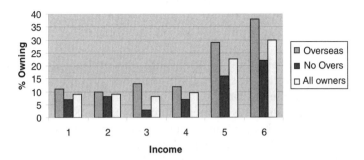

*Figure 6.2* Ownership of selected appliances comparing families with a member working outside of India (overseas) versus families with no family member working outside (no overs), by household income categories, based on a survey of 408 households in four Trivandrum neighbourhoods

networks. This pull of money and things through the 'scapes' of family is changing ideas about consumption and the ways commodities fit into consumption practices in Kerala.

## Dual residence

Early migrants from South Kerala were mainly from families in the lower economic strata. In their places of work abroad, they would find routine use of goods and technologies which were classified as luxuries in Kerala. This encounter is best expressed in the words of Gopal.

> Most of the people who have gone from here to the Gulf are not very affluent actually. They go there, make an earning. You have all these facilities. When somebody left from here he never had a refrigerator at home, he never had a TV at home, he never had an a.c. at home, he never had a car. These are all luxuries. But when you go you cannot live in that hot condition without an ac, and you cannot live without a refrigerator, that is a necessity. You have to have something cool (to drink) after a hard day's work. And, you need transport. Over there

petrol was not a concern. Petrol was cheaper than drinking water. So having a car was not a luxury but was more an easy conveyance, so most families used the car. And the houses over there are bigger houses. When they return they wish to have those things over here.

When Gopal talks about the work migrant's ways of classifying things as luxuries or necessities, he says that the confrontation with new ways of living disrupts those classifications. The habitus of everyday consumption in the residence abroad is different from that in the Kerala residence (Bourdieu 1984).[5] This confrontation of two routine ways of classifying and using goods lifts both into what Wilk calls the 'discursive sphere of heterodoxy...where, eventually, through the exercise of power, they (needs and wants or luxuries and necessities) can become re-established as orthodoxy, and eventually sink back into the accepted daily practice of the habitus (1999:10).' The dual residence of Kerala work migrants works this way, disrupting and reorganising the habitus of daily practice. Establishing a residence in the Gulf, in a home equipped with washing machines, air conditioners and cars, disrupts taken-for-granted classifications. As Cavita said about her home in Kuwait, 'All the modern things were there.' Had work migrants viewed their lives abroad as temporary, this disruption in ideas about luxury and necessity perhaps would not have any profound consequences for their consumption practices. However, Kerala work migrants see their dual residence as long term, perhaps even having a duration encompassing their entire working lives. Over time, living with them becomes a normal part of every day living.

Cavita's 'normalised' luxuries and the consumption practices in which they were embedded became taken-for-granted aspects of life after Cavita and her family retired in Trivandrum in 1990.[6] The confrontation with different routines was once again brought to light, this time when Gopal's new bride Kahena moved into the household.

Kahena followed the typical Kerala practice of cooking every meal from scratch, for the reasons discussed in Chapter 4: consuming foods that have been stored and reheated is thought to be unhealthy (leading to laziness and stupidity). Cavita found cooking from scratch every day to be a waste of time. Furthermore, the practice used more cooking fuel and was thus more costly. Cavita insisted that Kahina follow the practice that Cavita had established in Kuwait: preparing large amounts of food for several days of meals, keeping the food refrigerated and reheating small portions. In a conversation with all of the family members present, Kahina told me that this practice 'disgusted' her. Differences over food and refrigeration practices became a serious source of conflict that continued throughout my first year of contact with the family. It became so acute that Kahina was on the verge of moving back to her parent's home. When I visited the family a year later in 2003 they had reached a compromise. Kahina prepared some dishes from scratch, mainly for her own consumption, but she had gone along with Cavita's wish to cook most of the food in large quantities, store it and reheat it for weekday meals. Thus despite meeting resistance, a new consumption practice had gotten a foothold in this transnational household.

## Conclusion

Though initially sceptical about what I considered to be exaggerated claims about the contributions of migration to changing consumption, it became clear to me that the Kerala migration has been, and continues to be an important conduit for new commodities, and new consumption practices. The dimensions of migration have grown enormously, in terms of the numbers of people and families involved, and in terms of the amount of things and money drawn back into Kerala. Migration has touched nearly half of Trivandrum's population. Stereotypes about Gulf families live on and the sense of alienation has affected their consumption. Consumption is one avenue for work migrants to get from 'nowhere' to 'somewhere' in social space. However, from my observations and discussions with young people, I found that the practices of migration and of work migrants are, after 30 years of migration, on the way to achieving a sense of normality. Young males did not express the same concerns as their parental generation about being perceived as different.

We have seen how migration provides a conduit for the transfer of consumption practices from one home to another, and for changing what 'matters' in Trivandrum (Miller 1998). Goods and new practices move through the extended family and in this way extend the changes to sizable numbers of people. Gulf income gives many people the means to put new ideas about consumption into action. Life with a refrigerator, a stereo system, a television or an air conditioner, initially thought of as luxuries for the migrant, get normalised as means to achieve comfort, convenience, and entertainment. This reconfiguring of habitual practice through migration is an important impetus for changing consumption, as well as for the reconfiguring of ideas about the good life and the role of consumption in it. The next chapter takes up other important contributors to consumption change.

# 7
# Material, Discursive and Performative Contributions to Consumption

Thus far, consumption has been examined through the lens of social and cultural practices, including gender, caste, family and work. In this chapter, household consumption practices are in focus. How are changes implicated in the interaction between local practice and global agents, both commercial and developmental? I will examine how people keep themselves clean, stay cool and get around town. These three practices are chosen because they have a dense material component, in the form of household appliances and the energy, water and cleaning agents on which they are dependent. These are the 'inconspicuous' technologies that Warde and Shove have argued are so central to the ways people live their everyday lives, yet largely ignored by researchers precisely because they are unglamorous. The household appliances and the infrastructures into which they are a part (for example electricity and water) steer consumption in ways that will be explored in the chapter.

For products like soap and other cleaning agents, globalised ideas about development and progress are relevant to understanding consumption change in south India and elsewhere. After exploring the ways that soaps have been promoted in global discourses, attention will be turned to automobility, a 'conspicuous' commodity which is used by consumers as a form for social performance. The discussion will take up the rapid increase after 1991 of the kinds and types of automobiles available in India and how this has contributed the ways that the car is used in social performance.

## Keeping clean

Clean homes, clean bodies and clean clothing are all deeply important to Malayalee. The words for cleanliness and spiritual purity have the same root in Malayalam. *Sudhi* means clean and *parisudhi* can mean both 'very clean' and spiritually pure. As we saw in Chavita's typical day, outlined in Chapter 4, women put these ideas about home cleanliness into practice in their frequent rounds of sweeping, washing, and cleaning dishes throughout the day. The Kerala broom, made from palm leaves, seems to be in perpetual motion, as women pass it over floor, porch and garden many times a day. Cleanliness is important for people of all castes and religions, but Hindu doctrines are explicit about the relationship between cleanliness to spiritual purity. Our neighbour Sindhu (Hindu Nair), reflecting on this said, 'Just as the barber and the washer-man[1] take away pollutants from the body, the sweeper takes it away from the house.' In addition to rigorous daily sweeping, the Hindu family performs a simple *puja* (worship) every day. The daily lighting of oil lamps, candles and incense represents both a spiritual cleansing and an invitation to gods to enter the home.

In addition to daily cleansing, Hindu Nair and Ezhava families periodically ask a priest to conduct a ritual cleansing of the home (a cleansing *puja*). Osella and Osella (2000) write that until recently this was practiced mainly by elite Nair and Ezhava homes, but I found that people of all castes and socio-economic situations performed periodic cleansing of their home. The priest passes through all of the rooms of the house, praying and literally smoking out of spiritual impurities. For this, he uses a smouldering mix of pine sap, sandalwood powder, grains of gram, and rice. Though costly (our neighbours in Kumarapuram paid anywhere from 10,000–20,000 rupees for a cleansing *puja*), they are done by most families every few years, usually in conjunction with a major family event such as the birth of a child or a marriage.

The association of cleanliness with purity is also important to the way people clean their bodies. According to Gupta (2000:35), for Hindus, 'Substances that are routinely expelled from a person's body are polluting and dirty, even to the person concerned.' Sindhu explained that she always preceded her daily prayers with a bath: 'If we are clean and cool (from the bath), we can pray to the God more peacefully.' Everyone, young and old, bathes at least once daily and many bathe several times daily. Rigorous hand and face washing precedes and follows meals. Clothes are seldom worn more than once before they are washed. One need only take a stroll down MG Road, the main thoroughfare of Trivandrum, on a hot and dusty afternoon to be convinced of the importance of cleanliness. To the outsider's eye, people of all ages and castes seem to have just stepped out of the bath and into newly laundered clothing.

Interesting from a Western perspective, hot water is associated with good hygiene, but this is not the case in Kerala. Keeping the body and clothing clean is accomplished without the use of hot water and, until recently with only a moderate use of soaps and cleaning agents. Over the past decade, the use of soaps is increasing dramatically, but the use of hot water in cleaning is still rare. The next two sections will take up the reasons for stability and change in the consumption of hot water and soaps.

## Hot water consumption

Hot water is not seen as an essential ingredient in good hygiene, nor is it used to achieve bathing comfort by most people. A Syrian Christian male neighbour in his 30s told me, 'I prefer cold water. It is a habit. It is comfortable when I am normal (healthy). If I have some abnormalities like this pain or something like that I am forced to use hot water. That is my feeling.' The Malayalee attitude resembles that of the Japanese, who do not regard hot water as necessary to cleanliness (Wilhite *et al.* 2001). Unlike Malayalee, the Japanese do associate hot water with bathing pleasure (Wilhite *et al.* 2001; Benedict 1946). Most Malayalee in good health bathe in cold (tap) water. Water heating appliances in bathrooms are rare. Many people bathe at an outdoor water tap (dressed in an undergarment or *lungi*). Among the 408 survey participants, only 17 per cent of those under the age of 60 bathe in hot water.

Even though hot water is not a part of everyday bathing, hot water is seen as having therapeutic qualities, something that coincides with the principles of *ayurvedic* medicine. *Ayurvedic* diets and health remedies are very popular today in Kerala and elsewhere in India. The principles of *ayurvedic* can be traced back to ancient Hindu medical scripts and other documents dating from 2000 years ago (Wujastyk 1998). They involve a combination of diet, bathing in herbs and use of medicinal oils, as both preventive medicine and for curing various health problems. Kerala is known for its *ayurvedic* products and health centres. Most of the herbs used in *ayurvedic* can be found in Kerala. A*yurvedic* practices and products are especially popular among old people. Jeffrey (1992:189) found in the 1980s that ayurvedic medicine was practiced 'enthusiastically' by older people of all castes, and especially by members of the Ezhava caste. Several of our elderly neighbours followed an *ayurvedic* regime, which included frequent baths in very hot water, to which plants, herbs and oils had been added. Thirty-one per cent of the survey participants over 60 years old bathed regularly in hot water.

In Trivandrum and the coastal cities of South Kerala, many new a*yurvedic* centres catering to tourists feature hot-water baths, massages and herbal cures. Hot water is thus an important element in Kerala tourism, but is not important in the everyday lives of Malayalee. One reason is simply that hot water in the quantities needed for bathing is not available to most people. The technological infrastructures for water delivery and water heating are only available for small segments of the population. In order to commodify hot water, comprehensive changes would be necessary in electricity and water infrastructures.

By contrast, another component of hygiene, namely soaps, are some of the most heavily marketed products in Kerala (Bullis 1997). Both Indian and foreign-based companies, including Proctor and Gamble, Colgate Palmolive, and the Indian company Hindustan Lever have heavily advertised their products since the early 1990s. Malayalee are literally being overwhelmed by soaps, not only in commercial marketing but in the campaigns of development organisations.

## Soap as development

Hand washing with soap has been one of the recent subjects of World Bank programmes. As pointed out above, Trivandrum custom is to use plenty of water in the washing of hands, face and mouth, both before and after eating. A study in several Kerala *panchayat* in 2002 concluded that less than 20 per cent of the population washed with soap after using the toilet and less than 10 per cent before eating (Kurian 2002). At the same time, Kerala has achieved levels of hygiene comparable to European countries (Census of India 2001). This is reflected in Kerala's low infant mortality rate and low infant diarrhoea rates, by far the lowest in India (Franke and Chasin 1996). According to C. K. Soman (interview), leader of the NGO Health Action for People, good hygiene has been achieved without heavy soap use because of what he calls the Indian 'technology' for eating. The right hand is used for eating and the left hand for everything else. Babies and small children are not fed with utensils, but by the mother (her right hand). The differentiated uses of left hand/right hand accompanied by know-how about hygiene and access to plenty of water, makes excessive use of soap unnecessary in hand washing. Soap is mainly used after handling fish and other oily foods. More-over, Kerala's highly literate population understand the relationship between cleanliness and disease (Soman and Kuty 2004). Ironically, Kerala's high literacy is one reason why the World Bank and commercial marketers often choose Kerala as a testing ground for new products and programmes.

Valerie Curtis of the London School of Hygiene and Tropical Medicine received World Bank support to study how increased soap use can improve hygiene and decrease infant mortality and disease. She documented positive results from a pilot project in Latin America. Curtis, (quoted in Kurian 2002) wrote the following about the hygienic potential for soap in India:

> The soap industry might be doing more than the Ministry of Health to save lives in the country. Soap is probably one of the most important health products ever invented. Soap, when used to wash hands, prevents the transmission of the killer agents of disease and so prevents infection and death...The fact is that if everyone in the country washed their hands with soap after the toilet, deaths from diarrhoea could be halved, saving a thousand lives a day.

The World Bank accepted her arguments and gave their support to a hand washing campaign in India. The World Bank drew several multinational soap corporations into the project. The Chairman of Hindustan Lever, a subsidiary of Unilever, wrote that 'The aim of this programme is to change the lifestyles of millions of people across the country (*The Financial Express* 2002).' Five hundred million rupees were to be spent on the programme, half by Unilever and half by Indian national and local governments. *The Financial Express*, writing about the campaign reported that:

> The 'hand washing with soap' campaign will be done through advertising, doctors at primary healthcare centres and hospitals/clinics and *Aanganwaadis* as well as through posters and banners. Kerala has been chosen to kick-off the programme in India as it is the most literate state and chances of success in the state are high.

After the campaign was announced, it came under criticism from social and environmental activists. Soman was concerned for a number of reasons, first because adding soap to hand washing is costly, bringing an unnecessary household expense. In addition, the use of soap increases the amount of water needed in rinsing. This increases the demand on strapped water resources and increases the amount of grey water runoff into water bodies. Diminishing access to clean water is an acute problem in Kerala, contributing to problems with public water delivery and in

Kerala's many individual household wells, where ground water is increasingly contaminated. In Kannamoola in the early 2000s, water was only available a few hours a day, usually in the late evenings. Since most homes in this neighbourhood are not connected to public water lines, women waited with their buckets and containers for hours at a time for water to appear in the public taps.

Indian activists such as Vandana Shiva are critical of the World Bank's lack of concern for the social and environmental consequences of heavy soap use as well as for its neglect of local knowledge and achievements. Shiva (2002) wrote:

> Kerala has been chosen as the state to implement the 'Washing Hands' project in India even though Kerala has the highest hygiene standards, lowest diarrhoeal deaths, highest awareness on prevention of diarrhoeal diseases, lowest childhood mortality, highest female literacy... The World Bank project is an insult to Kerala's knowledge regarding health and hygiene. It is in fact Kerala from where cleanliness and hygiene should be exported to the rest of the world. People of Kerala do not need a World Bank loan for being taught cleanliness.

Partly as a result of the negative attention focused on the soap campaign, the World Bank project was first put on hold and then cancelled. However, the project illustrates how products like soap get uncritically promoted as development.

The association of soap with cleanliness has a long history. Vinikas (1989) outlines how 20th century US soap manufacturers invented new ways to underline the importance of soap in the face of declining use of soap at that time. In the early part of the century, the asphalting of roads and the replacement of coal-fired heaters with cleaner heating sources significantly reduced dust and ash. There was a resulting declining need for soap to clean dust and grime from bodies and clothes. In the absence of visible and tactile targets for soap, a new target – germs – was invented in a new cleanliness discourse (Barthes 1973). Germs were portrayed as being 'everywhere, omnipresent, ever ready to spread disease, debility, death (Vinikas 1989: 618).' Hands were portrayed as 'filthy organs' that needed to be washed frequently with soap. This slogan was adopted by national and local governments in the United States. Soap corporations were permitted to sponsor 'washing hands' campaigns in schools all over the country. This saga of soap and germs became common in discourses about cleanliness.

Comaroff and Comaroff (1992) reviewed colonial discourses and found that soap use was one of the important icons of progress and development in the British colonies. Soaps and cleanliness were subjects of a comprehensive colonial discourse (Stoler 1995). Burke (2003) relates how the producers of Lifebuoy soap, Lever Brothers, exploited this equation of soap with development in Africa. This use of soaps as an icon and as a political symbol continued into the era of development aid. Colloredo Mansfeld (1999:60) writes that in Latin America 'hygiene was a gendered and racial "micro-site" of political control.' Transnational soap corporations have been partners in many of the development programmes that have promoted soaps. Lever Brothers is one of these. They have drawn heavily on the soap-germ saga that was invented in the United States. In one oft-repeated Lifebuoy television commercial in Kerala, the opening shows a close-up of a man taking a bath. He is covered with germs creeping all over him and the bath tub. After using Lifebuoy, a new close-up shows that only a few limping germs have survived. Another frequently run Lifebuoy commercial is set at a zoo. A crowd gathers in front of an orang-utan, which is excitedly screeching and scratching its body. The people in the crowd in front of his cage are laughing. Then they realise that the orang-utan is mimicking a boy in the crowd. As the camera pans, we see that he is scratching rigorously. The crowd silences and looks at the boy and his mother with a mixture of pity and disgust. The boy's distressed mother hurries him away, buys Lifebuoy soap, and after a sudsy bath, we see that he has stopped scratching.

Lever Brothers was also the first soap company to emphasise odour in its advertising. It invented 'B. O.' (body odour) as another of the enemies to be vanquished by soap. Today, the 'fragrance market' is flourishing in India. Its marketing is all about reducing body smells and replacing them with commodified fragrances. According to a recent market analysis by the soap and cosmetic industry, 'The prestigious fragrance market, which has seen a steady influx of foreign brands...has an average annual growth rate of 15 per cent since 2001 (Phookan 2004:3).' Soap advertising in Kerala in 2004 was saturated by messages which associated good smells with being clean. Between the beginning of my field study in 2001 and end in 2004, I noted that the interest in fragrances had increased considerably. Perfumed soaps had grown in popularity and the use of perfumes was increasing. In a conversation about soaps and cosmetics in 2001, Chavita (whose typical day was outlined in Chapter 4) told me that she was not interested in perfumed products. When I returned to Trivandrum in 2004 after a year's absence, I noticed that she was liberally using perfumed soaps.

In recent years, there has been an expansion in the marketing and sale of ayurvedic soap products. The slogan of the popular ayurvedic soap, Jeeva is *'parisudhiyude parimalam'*, translated 'All purity is coordinated in this soap.' Thus the ayurvedic industry is capitalising on the association of purity with herbal products. Phookan (2004:5) wrote that 'In the last five/six years, there has been a renewed craze for herbal cosmetic and personal care products. Ayurvedic companies such as Fair & Lovely are producing soaps and their sales are growing rapidly.' Soap multinationals are also producing and heavily marketing herbal soaps. The soap multinational Unilever (through its subsidiary Hindustan Lever) is funding the local production and sales of what it designates 'natural soaps', using locally available ingredients. The soap is then sold door-to-door, reminiscent of some of the very early sales strategies that Western companies used to get footholds in Africa and New Guinea in the early 20$^{th}$ century (discussed by Wilk 2002b). Amway, a TNC which bases its operation around direct sales, is growing rapidly in Kerala. Its main products are soaps, cosmetics and cleaning products. Amway operates on the 'Tupperware Model' (Clark 1999). Local women are recruited as salespersons and they in turn recruit others. It is a pyramid system in which the women at the top of the sales pyramid get a percentage of the sales of everyone below them in the company sales hierarchy.

Hindustan Lever supplies the raw materials and the know-how for making soaps to members of local 'self-help groups', established under Kerala's comprehensive rural development programme of the 1990s (called the 'Peoples Plan Campaign', to be discussed in Chapter 8). Members of these groups sell the soaps at local fairs and from door-to-door. Another local development programme, the University of Kerala's *swadeshi* programme, supports local soap production and the direct sales of soap products. Twenty-five per cent of all soaps sold in Kerala in 2001 were produced with support from programmes that have 'local development' as their goal.[2]

To summarise, the related concepts of cleanliness and hygiene have been in a process of social reconstruction in which soaps and soap products have been equated with cleanliness and purity. Commercial companies and development organisations, both multinational and local are important change agents.

## Keeping cool

Another rapidly changing form of consumption is related to the practices of home cooling. Trivandrum has a hot and humid climate. The coolest period is December and January, when daytime temperatures reach the upper 20s (C) and evening temperatures seldom fall below 20 C. In Kerala summer temperatures rise to the mid-30s and the evenings are hot and humid. Because of the heat and humidity, an essential aspect of housing design and construction has been to foster a cool and comfortable living space. Prior to 1990, houses were cooled using a combination of natural cooling techniques (cooling accomplished without mechanical ventilation or cooling devices). These natural cooling practices are declining and the use of air conditioners is increasing. In this section I look at how historical changes in building practices, coupled with the onset of modern technological infrastructures have made the consumption of air conditioning inevitable. Since the mid-1990s, sales of air conditioners have been increasing, with a tripling of sales in Trivandrum in the three-year period 1999–2002. This change in cooling comfort involves changes in the social organisation of work, public building regulations, changes in the construction industry and consequent changes in the kinds of materials used in construction.

**Plate 8** AC advertisement, Kumarapuram, Trivandrum

## Changes in the design and materials in home construction

Houses of the 19th and early 20th century were constructed of mainly locally available organic materials. The construction of elite and upper caste homes was done by members of the *Viswakamos*, or artisan caste (Gopikuttan 1990:2085). The *Viswakamos* were the repositories of knowledge about construction, called *thachusasthram*. A house's design and physical orientation were decided upon in consultation with a priest or an astrological expert. Deviations from any of the principles of *thachusasthram* were believed to bring bad luck to the house and its occupants.[3]

*Thachusasthram* encompasses virtually everything related to the design and construction of the house, including selection and orientation of both the plot and the building itself, placement of windows and doors, shading and ventilation, and timing of the occupation of the home. The materials used were locally available and mostly organic, which promoted air flow in and out. Wood was the main building material for upper caste Brahmin and Nair houses. Middle and lower caste houses were constructed of mud, unburned bricks, laterite, bamboo, straw and leaves (Harilal 1986).

By the end of the 19th century, the role of the *Viswakamos* in construction began to decline. There were several interrelated developments which contributed to their diminishing role. First, Indian elites in the 19th century colonial period shifted their interest towards Anglo-Indian housing designs. From the point of view of cooling, the colonial designs also incorporated principles for natural cooling. Porches, slant roofs and the use of tile roofing materials all promoted air flow. Another development from the mid-20th century was the adoption of capitalist principles such as profit, mass design, and segmentation of tasks. New public laws and regulations demanded blueprints, cost assessments and the submission of written plans for new buildings, as well as site plans and fee breakdowns, all of which had to be certified by architects and engineers. Government loans were authorised only on the recommendations of formally trained construction engineers. The growing construction industry recruited and supervised workers (Harilal and Andrews 2000). Unskilled labour was used whenever possible. Less costly, pre-fabricated materials from other parts of India (and abroad) replaced local materials, both in public buildings and in house construction. These developments all contributed to the elimination of traditional artisans from building construction (Gopikuttan 1990). The principles of natural cooling were eliminated with them.

Today, the principle materials used in housing construction are cement, plaster (laterite) and burnt bricks, all of which have poor thermal properties in hot and humid climates. Among households participating in my Trivandrum survey, over 84 per cent had homes that used combinations of these materials. Roofs constructions today are made mainly with cement. After the 1930s, concrete began to be used in roof construction. Traditional roofs were thatched with coconut leaves and straw. Clay tiles began to replace thatched roofs in the mid-19th century. Both thatch and roofs made of clay tile allow air flow, yet inhibit the penetration of thermal radiation, both of which are excellent features in hot and humid climates. Concrete is cheaper, lasts longer and does not require much maintenance; however, concrete is a very poor thermal performer because it collects and transmits thermal radiation into the house, and then traps the heat inside. By 1941, 48 per cent of houses in Travancore (the most southerly of the kingdoms subsumed into Kerala) had concrete roofs. By 1960, 75 per cent of Trivandrum houses had concrete roofs (Gopikuttan 1990). Eighty-four per cent of participant's homes in the Trivandrum survey in 2002 had concrete roofs.

Aside from its low cost and structural advantages, the price of concrete declined rapidly between 1970 and 1985 as production increased (Gopikuttan 1990). The coming of electricity and plumbing favoured concrete in wall construction, and as kitchens and bathrooms moved into the house, the need for metals, plastics, pipes and wiring in the walls and floors of houses increased. Today, the concrete industry is booming in India, with production increasing in recent years at about 10 per cent per year (and parenthetically, producing huge amounts of the climate gas $CO_2$ which is a chemical by-product of concrete production).

Another development affecting thermal cooling is that house sizes increased rapidly from the 1970s. Work migrants contributed to this.[4] According to Jeffrey (1992:205), from the early 20th century, work migrants attempted to 'turn their money into status by building imposing houses, out of keeping with their former place in society...Like costume, houses symbolised the social order of the old Kerala'. Sindu (a Hindu Nair) told me that 'in the olden days having an elephant was status. For Gulf families, having a big house, that is status'. As discussed in the last chapter, accruing or showing status may have been a motive for some, but many were simply replicating ideas about house size and design picked up in their places of work abroad. I asked 24-year old Rija, who is looking for a husband and whose father and uncles all live and work abroad to describe her 'ideal' house. She said that she did not want a huge house, because it would take a huge effort to clean and maintain.

## Material, Discursive and Performative Contributions to Consumption 115

She wanted a house that is 'modest' in size. However, when I asked her to describe it, she said that it should have two sitting (living) rooms (one for everyday use and one for guests), two kitchens (modern and *chula*), and a bathroom large enough to room a bath/shower. A house big enough to incorporate these spaces would be considerably larger than her grandfather's home (her home for the six years she has been at university in Trivandrum). Thus the 'modest' house for this young woman who grew up with her parents in north India is much bigger than her grandfather's home from the 1950s.

Bigger houses have a greater volume of air to cool, and either more air conditioners or bigger compressors. In either case, the result is an increase in the amount of energy and cost involved in cooling homes. Another house design that is untypical for Kerala and is contributing to both larger volume and the need for air conditioning is the two-storey house (see Plate 9). Two-storey houses are constructed using concrete because it is cheap and strong enough to support the added weight of the second floor. The air conditioning demand of these structures was brought home to my family after living in one of these two-storey, concrete houses for four months in 2003. The heat gain throughout the day made the second storey virtually unusable without the room air conditioners on constantly and at full force. This contrasted starkly with the comfort of the house we occupied during our first two visits to Kerala. It had been built in the 1950s. It incorporated

**Plate 9** Gulf house, Kawdiar, Trivandrum

some elements of Kerala architectural style of that decade, combining elements of traditional Kerala and colonial inspired architecture. It was designed for living without air conditioning, featuring a single storey, a long corridor through the middle of the house (facing in the direction of the prevailing breeze), a porch around the perimeter with a continuous band of windows facing onto the porch, a marble floor and a tile roof. The natural cooling principles incorporated into the design of the house and into its building materials made the house fairly comfortable even in the warmest periods of the year.

Architect Joseph expresses the change in house styles and sizes from the 1970s as a 'tragedy':

> The real tragedy of Kerala happened maybe 30 years back when suddenly these architectural companies came and people didn't really strike the right balance. They just went around making all of these monstrosities... We have to live with them. And I think it is a very nice reminder of what mistakes can be like...In these post 1970s structures, you cannot stay in a room for more than 10 minutes (because they are too hot).

Interestingly, today Joseph exclusively designs and builds houses for middle class and elite families that are meant to have an air conditioner. He delivers finished houses with air conditioners installed.

Changes in the urban environment over the past few decades have also contributed to the consumption of air conditioning. The density of buildings has increased in Trivandrum as agricultural space within the city limits has been sold and developed. In 1980, for example, 10 per cent of the area of Trivandrum was rice paddy. Today there is no rice paddy within the city limits. As the population has grown and joint family households have declined in number, the density of housing has increased and the amount of vegetation has declined. As a result, breezes and shading have both been reduced. Another problem is increasing air pollution, mainly from the increase in vehicle traffic. Many people keep their windows closed in order to block out pollution (Sharma and Roychowdhury 1996); one of my neighbours referred to his newly purchased air conditioner as an 'air purifier'. Finally, there has been an increase in the number of mosquitoes, mainly due to poor water runoff and the stagnation of rivers choked with garbage. People close their windows to keep mosquitoes out. This also closes out fresh air and makes air conditioning more attractive.

## Changes in the political economy of air conditioners

The data provided to me by air conditioner distributors and retail sales managers, show that sales of air conditioners tripled in the three-year period from fiscal year 1999/2000 to 2001/2002. Data from earlier years is not reliable, but it is clear from all-India data that 1991 marked a watershed for air conditioning; it was the same year that India dropped or modified restrictions on foreign products and foreign investment. Prior to 1991, consumer-goods imports were permitted only with a special import license or on the basis of proven need. In 1991, there were only three or four Indian manufacturers selling air conditioners, and there were almost no showrooms or retail sales. To buy an air conditioner, one had to contact a dealer and place an order. This changed rapidly over the following decade. By 2002 there were more than 17 major air conditioning brands on the market, most of them foreign (some with licensed production in India). There are also a number of small new Kerala companies making air conditioners, or air conditioner components, supported by a programme of government grants to small businesses. Altogether, taking brand and model into account, there were about 60 different air conditioning models available in Trivandrum in 2002.

Both competition and new tax policy lead to a gradual drop in prices between 1991 and 1996. Nationally, an air conditioner of average capacity (1.5 ton compressor) sold for an average price of 32,000 rupees in January 1993; by January 1996, its price had fallen to 27,000 rupees. The price fell further after 1997, when the excise tax and other import restrictions on air conditioners were eliminated, and then again in 2001 when air conditioners were removed from the luxury goods category, thus eliminating the 18.75 per cent luxury tax. Altogether, retail prices of air conditioners have fallen about 20 per cent since the mid-1990s. Dealers told me that stiff competition among brands and dealerships means that customers can negotiate a price below the list price. Also, the interest rates on product loans have decreased. Some financial institutions require only a small down payment and even offer interest free payments over a 24-month period. Lower prices and lower monthly payments make air conditioners affordable for many middle income families. According to the sales managers in Trivandrum appliance stores, their core market earlier consisted of wealthy families. From the mid-1990s their sales focus has been on the two-income middle class family.

Dealer statistics do not capture all of the air conditioning sales. There is a considerable trade in illegally imported air conditioners. Malayalees can easily cross the border into Tamil Nadu, where there is an active trade in smuggled goods of all kinds. There, an air conditioner can be purchased 10–20 per cent under the Trivandrum retail price.[5] Work migrants are also a source of air conditioners. They are allowed to bring back appliances duty free if they can prove they have used them for a minimum of one year in the country in which they work.

People who have air conditioners use them sparingly, partly due to the expense of the electricity needed to run them, but also because voltage is unstable, and without an expensive voltage regulator, the air conditioner's compressor can be damaged by excessive use. Most Trivandrum owners only turned their air conditioners on for periods of two to four hours a day, though some let them run throughout the night. In homes like that of Syrian Christian Abraham, whose parents share the home with him and his wife Maya, all of the bedrooms have room air conditioners. Their practice is to use the bedroom air conditioners for their afternoon siesta and while sleeping at night. Very few families use their living room air conditioners during the daytime or in the early evening. My impression is that the cost and quality of electricity are the main reasons why those who own air conditioners are not using them more extensively. As Joseph told me:

> I personally feel there is definitely an upwards surge in the use of air conditioning. The only thing that is probably impeding it, I mean anybody can make a one time investment in an ac, but then when you are also looking at the power bill which is coming every month for 3000 rupees, that is probably the main point which is holding back air conditioning.

### Alternatives to air conditioning

Why do people not seek or construct homes that can be cooled passively? First, in established neighbourhoods such as Kumarapuram, there remain very few houses that were designed and built to be lived in without air conditioning. For those who have the resources to build a new house, there are very few building entrepreneurs or architects who routinely build houses amenable to passive cooling. Examining the curriculum of Kerala's most prestigious Architectural Department at the University of Kerala's School of Engineering, neither the concept of passive design nor any practical courses on tradtional architectual principles exist. This is noteworthy, given that one of India's best known architects of passively designed houses is from Kerala, Laurie Baker. Baker designs and builds houses that encourage the use of natural cooling. From the 1980s Baker put designs into practice in Kerala housing, building a number of them in Trivandrum for upper middle class and wealthy families. His house designs use unfired bricks, limestone mortar and arched doors and windows. His arched design is used to distribute the weight of the structure and allow for more porous cement.

Baker achieved Indian and world wide attention when he designed and built the buildings that would house the Centre for Development Studies (CDS) in Trivandrum, one of Kerala's important universities. He used local materials, unfired bricks and light cement, as well as incorporating arched windows and doors. Some of the staff of CDS hired Baker to build their own houses. Interest in the 'Baker house' peaked in the 1980s but has declined since. The Baker design did not take hold as an attractive middle class design, despite Baker's widespread reputation in Kerala and elsewhere in India.

I found that one reason for the lack of interest is that people find that the 'Baker house' looks unfinished. This is because Malayalees typically finish their houses with a layer of plaster. Baker insists that his brick structure not be plastered or painted because this would inhibit ventilation (or as he put it 'the way the house breathes'). Dissatisfaction with the design came clear in discussions with low-income families who were the recipients of government grants to build new houses designed by Baker and his colleagues. Many people complained about the design, especially the unfinished look. People told me they would rather use the government grant to 'finish' their house (meaning to add a plaster façade) than using the money to finance Baker's naturally cooled design. The fact that Baker's design is neither traditional nor modern may explain why more people are not interested in it.

### The normalising of air conditioning

To summarise the discussion of home cooling, changes in availability of materials, costs of alternatives and new ideas about what constitutes a modern house have paved the way for air conditioning. The changes involve a change in local building technologies (knowledge, designs and materials); in public regulation; in the spread of capitalist business principles in construction; and in the use of mass-produced materials. In short, this change in consumption is not the result of a latent need waiting for the right technology at an affordable price. It is rather the result of a socio-material change in the home as technology, new ideas about how to achieve comfort from returning worker migrants, and increased availability and affordability of air conditioning. The most important point is that even the most tradition-minded, nature-oriented or environmentally-concerned family would have difficulty living in a modern Trivandrum home without air conditioning.

Today, living spaces which have been made unliveable through a half century of changes in house designs can be recovered by the air conditioner. However, as has been pointed out, the increasing use of air conditioning is contributing to Kerala's energy crisis. The increasing use of the air conditioner is one reason for expanding residential electricity consumption. Residential energy use grew from16 per cent of Kerala's total energy consumption in 1995 to 44 per cent in 2001. As a result, power blackouts, both planned and unplanned are increasing. Further, the electricity consumed for air conditioning by middle class and wealthy families diminishes access to electricity for poor and lower middle income families.

Interviews with air conditioner owners reveal that once installed in the home, air conditioning changes the ways in which people use the house as social arena. Air conditioning affects the ways families move and interact, both within the home and without. Demand for a closed, tight structure requires that doors and windows must be kept closed to keep cool air inside. Those who had installed air conditioners in the living room usually kept the entrance door from the street closed, as opposed to keeping it open or ajar as is the usual practice in Trivandrum homes. This reduces the movement of people in and out of the house and reduces the family's social interaction with neighbours and others moving along the street. The front porch in air-conditioned homes is used little or not at all.

These changes resemble changes in other countries where air conditioning has become common, such as the southern United States and Japan (Holston 1999; Cooper 1998; Wilhite *et al.* 1997). Much of social life takes place in enclosed and conditioned spaces. Air conditioning has brought with it enormous transformations in the ways people move, dress and in the case of Japan, undress. In Tokyo, the electricity demand created by air conditioning is so acute that people are being asked to shut down air conditioning and suffer through several hours a day in hot buildings made of concrete, glass and steel that no one imagined would ever be used without air conditioning. In the Southern United States, virtually every aspect of everyday life takes place in an air conditioned environment. On a typical day in Orlando or Houston or Phoenix, people leave their air conditioned house in the morning, gets into their air conditioned car and drive to their air conditioned office. After work they shop in air conditioned shopping malls.

Movement outside these air conditioned corridors is diminishing. Sidewalks in Southern US cities are being removed because no one walks anywhere anymore.

In Kerala, I was frequently told by non-owners of air conditioners that air conditioner owners had bought them in order to show off. The owners themselves consistently denied this, claiming comfort as their motive. I only encountered one owner who unabashedly consumed new appliances in order to cultivate an image of being a social innovator. This Hindu Nair man in his 50s bragged about the new 'gadgets' he had purchased for his home, showing me his foreign cooking stove, expensive stereo system and his central air conditioner. He proudly related how his brother had recently told him, 'What I am thinking now, you started thinking 20 or 30 years back. What I am seeing in my children, and what I have accepted now as normal, you started doing 20–30 years back.' This man obviously sees his consumption of modern appliances as a kind of social performance. In my view the performance motive is minimal for the acquisition of the air conditioner and other 'inconspicuous' appliances (Shove 2003). However, it is highly relevant for understanding changing mobility, the subject of the next section.

## Modernity on wheels

In the course of a single decade, cars have become important icons of progress for both the Indian nation and middle class Indian families (the contributors to Miller 2001b claim that cars are icons throughout the developing world). An article in *The Hindu* captures both the macro-economic and personal importance of cars (Rajashree 2001).

> When we analyse the data of motor vehicle population registered in Kerala since 1980, it has become quite clear that the rate of growth is more than 900 per cent till 2000. No doubt, this stupendous growth rate …targets the needs of the day to day life style of the modern citizens of our Country. The concept of utility of motor vehicles has undergone a sea change, especially after the advent of the new Millennium. Owning a motor vehicle in this modern age has become a necessity rather than a luxury, as we have believed in the past. Now-a-days, people from all strata of life … need at least one type of motor vehicle.

Cars are depicted as both modern and necessary, a double thrust that bears a powerful agency for change in Kerala. Cars are moving into families and onto roads at a remarkable pace. One of my greatest surprises on arriving in Trivandrum in 2001, after a 10-year absence from India, was the density and variety of vehicles on the road. In our neighbourhood of Kumarapuram, there seemed to be a car in virtually every driveway (the survey of middle class neighbourhoods later showed that about 40 per cent of the households owned a car). Interviews with sales managers of automobile dealerships revealed that sales of foreign models were increasing at about 9 per cent yearly and that sales of Indian models, the Maruti and Tata, were increasing at about 5 per cent yearly. This growth in car ownership is supported by data from the Indian Department of Transport (2002), which shows that the number of automobiles in Kerala increased by 50 per cent from 1990 to 1995 and then almost doubled from 1995 to 2001, a slightly higher growth rate than that for India as a whole. There is no doubt that the car is replacing the motorcycle as the preferred means of middle class mobility.

For those who can afford a car, it offers many practical advantages over the motorcycle, three-wheelers (auto rickshaws) and buses. For the car owner, the choices of when, why and for how long she or he travels are no longer subject to the vagaries of public transportation. Auto-rickshaws are sometimes hard to find in residential neighbourhoods, and their open design leaves the passenger exposed to heat, dust and pollution. Buses are slow, overcrowded and hot. The motorcycle and bicycle expose the driver to heat, dust and pollution, and in the monsoon season, to the frequent deluges of rain. The car has the advantage of accommodating the whole family (though one still sees families of up to five driving around town on their motorcycle). The disadvantages of using a car in Kerala seem to be totally disregarded by most car owners. These include congested, narrow roadways; speeding trucks and buses that regularly use oncoming lanes; and both pedestrians and bicycles, who, not having a designated space of their own, occasionally wander into traffic. Amongst our neighbours who owned a car, these kinds of problems did not seem to inhibit anyone from using their car for commuting, shopping, visits to members of the extended family, and weekend outings.

Social performance is clearly an element in car consumption. My Syrian Christian friend Abraham told me, 'The easiest way to show off your wealth is to buy a brand new car. It is the most identifiable object of assessment.' The car is effective in 'showing off' because it is visible, either parked in front of the house or moving around town. When I asked an owner of an expensive Toyota Sports Utility Vehicle (SUV) what motivated him to buy the car, he answered, 'They first look at the car, *then they look at me.*' He wanted people to see him as rich enough and savvy enough to own a car that has assumed a place at the top of the automobile hierarchy.

The SUV and other new luxury models have disturbed the long stable social hierarchy of car ownership. Prior to the 1990s, the Ambassador and a smattering of other brands were the only cars on the road. The Ambassador was intended for, and driven by the rich and elites. In the 1970s, Sanjay Gandhi, the son of then Prime Minister Indira Gandhi promoted the idea of domestic production of an Indian 'everyman's' car. The idea was that it should be affordable for the middle class, should have a practical design, maintained from year to year so that it could be cheaply and easily maintained. The first Maruti, designed and produced in cooperation with the Japanese company, Suzuki, was produced in the mid-1980s. For the unaccustomed eye, it is difficult to differentiate between a Maruti from the late 1980s and one produced a decade later.

The Ambassador and the Maruti provided a fixed and stable social hierarchy. With some few exceptions, in urban Trivandrum the rich and elite drove Ambassadors, and most everyone else able to afford a car drove a Maruti. This began to change rapidly after 1991 and the opening of India to foreign models, manufacturers and investment. New foreign luxury cars such as the SUV, very popular in North America and Europe, competed with the Ambassador for ranking as the highest status vehicle. By 2001, SUVs had outstripped the Ambassador in status. One need only observe the pecking order on the road to confirm this. SUVs are given free reign to use oncoming traffic lanes and all other vehicles, with the exception of buses and police jeeps are expected to make way.

The Maruti also got competition in the 1990s from both foreign models and from Indian manufacturers. The Korean companies Hyundai and Daewoo introduced cars priced only slightly above the Maruti. In the mid-1990s the Indian corporation Tata, India's largest producer of trucks, buses and other heavy vehicles, began producing the Indigo, with a similar design and competitive price to the Maruti. The manufacturer of Maruti has itself begun manufacturing a more expensive and sportier model, the 'Zen'. The standard Maruti is still selling well, but one indication that its popularity is waning is the fall in price for a used Maruti in 2001 and 2002. Reduced import taxes, low-interest loans and long payback times made more expensive cars affordable for middle income families. This expanded and more articulated field of vehicle types, looks and prices, ranging from sedans to elite vehicles provides a much broader range of potential social statements for their owners.

Architect Joseph described to me how he uses cars to classify his clients (along the lines theorised by Douglas and Isherwood 1979 who wrote that people use goods in cognitive classifications). Joseph 'gauges' people by the type of car they drive and uses the classification to decide which type of house design to sell them:

> How do you really gauge style? If he (a potential client) would come in a very flashy red car, then you know he is a neo-rich guy who is struggling to find the right balance. The other guy who comes in a grey or black car, he is a sober guy, who wants luxury, but doesn't like to be very brightly noticed. So you have to please him. He might be the type who had to be forced into the back seat of a Lancer or a big luxury car, probably because his son insisted for him to buy a big car. So he will be a reluctant customer. He will be wearing his *lungi*. He can very well afford properties and houses but he doesn't want to show it. So my business means that I simply have to observe. That is the way I can have a very satisfied customer. The flashy guy is satisfied, the sober guy is satisfied, the classy guy is also satisfied.

Another change in mobility consumption is that the status accrued from longevity and good maintenance is waning. With the Ambassador and Maruti, maintaining the car so that it looked and functioned like new used to be a source of pride. Today, having the latest model is more important. This change is very similar to that reported by O'Dougherty (2002) in her study of middle class Brazilians. She writes that in the past people placed importance on upkeep. Observing people's cars in middle class neighbourhoods, she wrote that she had 'the feeling that most (used) cars are new'. However, she found that increasingly, high status accrues to owners of a new car, called *carro zero* (car with zero kilometres). 'Not having a fairly new car (not to mention not having one at all, which for middle-class Paulistanos is unthinkable) can be a source of embarrassment (2002:45).'

This change from equating status with perfect maintenance to equating it with ownership of the latest model fits with Appadurai's (1996) distinction between consumption and consumerism. One important difference is that consumerism involves an acceleration in the pace of demand for the newest and latest object. Bauman (interviewed in Rojek 2004) goes further, associating consumerism with 'infinite desire.' Consumerism is equated with 'liquid modernity,' in which 'the meaning of the material or substances purchased become (sic) completely overshadowed by the signs and symbols of possessing and using (2004:291).' From what I have observed in Kerala, car consumption is still far from qualifying as consumerist in either of these conceptualisations. However, the changes of the 1990s resemble the first steps towards car consumerism in North America half a century ago. Examining marketing images in both places, in pre-World War II North America cars were advertised as 'built to last'. The car was promoted as the 'family car'. Compare this with marketing in India of the 1990s, when 'The four members of the state-sponsored family model (man, woman, son and daughter) have been replaced by man, woman, child and car (Fernandes 2000:623).'

In the US during the 1950s, there was a transition from associating the car with family to identifying it with its individual owner. Since the ideal family had two adults in it (husband and wife), the 'two car' family became synonymous with the ideal family. Today in the USA, not only do both adults have their own car, but teenagers in many middle class families expect that a car will accompany their first driver's licence (Wilhite and Lutzenhiser 1998).

India is still in the one car per family stage of automobility; however, among the wealthier families in the upper middle class neighbourhood of Kawdiar in Trivandrum, cars are becoming normal possessions for their teenaged sons. I discussed with Hindu Nair Rema why she and her husband had purchased a car for their 19-year old son: 'He is prone to all these things. Things which give status. He wanted a flashy car. We understand that he needs it. He is trying to be one of the crowd. I have given in to my son. Our society is very rigid, I think you must have found that out by now. We go by these strict rules. It is terrible in fact.' Note that Rema begins by talking about status, but ends by attributing the purchase of her son's car to the pressures of 'being one of the crowd'. Something similar was expressed by Syrian Christian Thomas when I asked him whether status was a motive for consumption of cars and household appliances: 'We know what other people are using. And so many changes are coming. I feel we have to move along with it.' Neither Rema (and her son) nor Thomas wants to be identified as being left behind, with its connotations of being different, out of touch or too poor to have what everyone else has. Rema and Thomas unwittingly reveal a subtle dynamic important to consumption: consumption is increasing in tact with the tempo of the push to get ahead and the pull to keep up.

## Conclusion

In South India, the ideas of what constitutes a clean body, cool house or good means of transportation, and the products thought to be essential to achieving these are changing. New ideas about cleanliness are wrapped up in a discourse of development, or progress. Development agencies, the governments of Kerala and India, and commercial actors have participated in the promotion of this discourse. Advertising is one of the outlets. There was a watershed in the 1990s during which the volume of advertising increased astronomically (I return to a discussion on the importance of advertising in Chapter 9). Soaps and other cleaning products are heavily promoted as 'necessary' to achieving a clean and hygienic body and house. Not only is soap portrayed as essential, but new constituents have been added to cleanliness which open for new forms of commoditisation. Smells, oils and natural ingredients allow people to 'customise' their bodies and appearances. As Shove (2003:103) puts it: 'having stripped the body of its natural odours people are free to customise aspects of themselves (skin, hair and scent) in any way they want'. In this process, the ideas, indicators and material constituents of cleanliness are being reconstructed in ways that encourages changes in consumption.

Concerning home cooling, we have seen that changes in the material constituents of house and the creation of the home are highly relevant to change. The social organisation of building, as well as the house's design, composition and infrastructural connections have changed, gradually but inexorably, similar to earlier changes in the United States (Cooper 1998; Kempton and Lutzenhiser 1992), Japan (Wilhite, Nakagami and Murakoshi 1997), and China (Wilhite and Norgard 2004). These inter-linking socio-material systems are changing and bringing with them an embedded demand for air conditioning. Technology, even on the grand scale of electricity infrastructures is not deterministic; however, it creates potentials for change, and therefore ought to be included within the scope of household consumption studies (see also Spaargaren 2004; Rip and Kemp 1998 and the contributors to Southerton, Chappels and Van Vliet 2004 for an exploration of the importance of technology infrastructures in everyday consumption).

The conclusion from this study in Kerala is that the material world is not only a consequence of earlier consumption but is itself a change agent. And since the material constituents of the home are increasingly interwoven with technologies, technology ought to get more attention in anthropology. Anthropology's principle subject, the 'other' represented by the non-western, rural and non-elite peoples of the world, has not possessed household appliance technologies in significant numbers, at least not until recently, to merit attention. Based on the changes I have outlined in south India thus far, this is clearly no longer the case. Technologies and other products of Western materialism can no longer be written off as somehow exogenous to local worlds and ignored. Another problem for the anthropologist is that while people are acquiring new products at a rapid rate, the changes they make in practices require a longer time perspective than the usual one or two years of observation can capture. An accounting for the 'endogenous histories' (Comaroff and Comaroff 1992) of local practices is essential to ethnographic enquiry about consumption.

Car consumption has a different dynamic. Consumption of automobiles changed dramatically after 1991 and the 'opening' of India to global markets and investment. Significant changes in kinds and varieties of cars – supported by changes in laws regarding imports, foreign collaborations and financing – coupled with a latent interest in products of foreign origin, have provided the raw materials that support the use of cars in social performance. The possible number of points in Bourdieu's social space (1984) for automobility has expanded dramatically over the course of a decade. People are using cars because they are convenient, provide flexibility and provide freedom, but also to make social statements about themselves and their families.

# 8
# Frictionless Political and Religious Ideologies

The interpretation of changing consumption has thus far concentrated on social relations, practices and development discourses. This chapter addresses political and religious ideologies that either directly or indirectly emphasise thrift, frugality or moderation in people's pursuit of material goods. In Kerala these have been embedded in India's post-independence political posture *vis-à-vis* the West, in Kerala's socialist political orientation, and in the religions of Kerala. Each of these are explored in the chapter, with attention given to the question of why they have not provided friction to the increasing interest in material goods and consumption.

The chapter begins with a discussion of Gandhi and his ideals of 'indigenous development' and material simplicity. These were rooted deeply in both Indian politics and in individual Indian identities for several decades after Independence, but while Gandhi and many of his ideas are still venerated, his ideas on thrift and simplicity began to lose their force after the 1980s. From the 1990s, consumption has been openly embraced in both national and personal development. In the second section of the chapter, I address the seeming anomaly that one of India's highest consuming states is also one dominated by a history of democratically elected socialist governments. The middle class has participated extensively in socialist projects yet consumption of public services has declined and private consumption has increased dramatically over the past two decades. Finally, religiosity in Kerala is discussed and the reasons why increasing interest in religion has not dampened interest in consumption.

## Gandhi's legacy

Gandhi's legacy is not something that I assumed beforehand would be relevant to everyday life in Kerala in the first decade of the 21$^{st}$ century.

However, from my very first meetings with neighbours in Trivandrum, the subject of Gandhi was brought up by people in virtually every discussion about consumption and social change. Gandhi was proposed as being relevant to India's efforts to alleviate poverty, justly distribute wealth, promote religious harmony, and foster non-violent change. However, Gandhi's ideas on material simplicity and scepticism to Western materialism, so glaringly important in the decades after Independence, were no longer seen as relevant to the ways people conduct their everyday lives. In this and the next section, Gandhi's ideas about consumption in national policy and for individual lives are explored, as well as the reasons behind the decline of their importance as references for consumption.

Gandhi is undoubtedly the most important icon of India's independence movement and one of the most important contributors to India's post-colonial political platform. Gandhi's political and social vision for post-independent India both had their roots in his and India's struggle to free itself of the British colonial authority. The Congress party, which Gandhi helped to found, was identified by Indians as *the* independence party by peoples across India. Malabar was under the colonial authority, and though the south-western parts of Kerala were not directly under colonial rule, the struggle for Independence was fierce in Cochin and Travancore.[1] Gandhi visited southwest India many times in the 1920s and 1930s in order to promote his political and personal vision. His popularity was evidenced in the extensive popularity of his clothing style (Jeffrey 1992). He had invented a simple dress consisting of cloth *dhoti* and *lungi*. He made a point of making his own clothing on the traditional Indian spinning wheel (Tarlo 1996). He designed and made a cap from local materials. According to Jeffrey (1992), Gandhi's dress was adopted by many young Kerala men in the 1930s. Not only did the style become popular, but many men began spinning their own *lungi*. The wearing of the *lungi* became a political statement, identifying the wearer as a supporter of the independence movement. Jeffrey tells the story that in 1938, many Congress party members arrived at the Kerala State Assembly dressed in *lungi* for the first time. This drew wild support from the assembled crowds. Those legislators who were not wearing a *lungi* scrambled to find a Gandhi cap in order to identify themselves with Gandhi (Jeffrey 1992).

Gandhi's political ideology was multifaceted. It included his important principles of *ahimsa* (truth and non-violence) and *swadeshi* (indigenous development). *Ahisma* is perhaps the most widely known of his principles outside India, but it was not universally accepted in the resistance movement as the appropriate strategy for throwing off the British rule. Even within Gandhi's own Congress party, there was strong disagreement about *ahimsa*. The left-leaning factions in the movement advocated armed violence. In Malabar, disagreement over the use of violence was behind the split in the Malabar Congress party, one faction breaking out and becoming the Kerala unit of the All India Congress Socialist Party – later the Communist Party of India (CPI).

*Swadeshi*, on the other hand was widely supported. *Swadeshi* emphasised local development, local knowledge and frugality. It characterised Western values and Western development as a threat to true Indian development. Implicit to *swadeshi* was the proposition that India would only truly develop, both spiritually and materially if the Western development paradigm and its agents were held at bay. India should determine its own future, drawing on its rich religious and agricultural heritage (Narayan 1968).

For Gandhi, the most important Indian development tasks were the extension of social and economic benefits to the poor. This would be accomplished not by wiping over India's indigenous skills and traditions and replacing them with modern technologies, but by building on them. Development of the Indian nation would come by way of 'the handiwork of each individual Indian, in their quotidian (sic) habits of production and consumption (Khilnani 1998:73)'. Simplicity and moderation were seen as essential to achieving fulfilling lives (Corbridge and Harriss 2000). As Mazzarella (2003:6) expresses it, 'sublimating these (material) desires for the greater good of the nation became, for Gandhi, every Indian's first duty'. Varma writes that *swadeshi* was associated with 'A reticence towards ostentatious display of wealth, which was seen as something in bad taste and incongruent in a country as poor as India (1998:32).' This was partly due to Gandhi's influence, but can also be traced to the Victorian valuation of thrift, one of the legacies of the Victorian British colonisers. According to Jeffrey (1992) and the accounts of the older people interviewed, thriftiness and frugality were important social norms for the Kerala middle class in the years after Indian independence. As we will see, this began to change in the 1980s and 1990s. The importance of *swadeshi* as a national political strategy declined during those years, although it would be resurrected in the Hindu nationalist politics of the late 1990s. In the next section I trace this fall and resurrection of the importance of *swadeshi*.

## The fall and rise of *swadeshi*

Jawaharlal Nehru, India's first prime minister and the man who is credited with 'inventing' the new Indian State, shared many of Gandhi's ideas, but there were subtle differences in their respective ideas about Indian development. Concerning India's relationship to the West, what Nehru called his 'third way' did not imply a total rejection of Western influence or economic assistance. India under Nehru accepted considerable amounts of development assistance from the United States, European countries and the Soviet Union. Nonetheless, India was to be in control of its own fate, expressed in the concept of 'self reliance' *(swaraj)*. Varma (1998:32) summarises the political principle of 'self reliance' this way:

> ...optimism in India's intrinsic economic strength and the political need to be insulated from external manipulation ... Self-reliance, both in the economic sense and in the political sense of being resistant to external pressure, was accepted as a valid policy paradigm to ensure that independent India would be able to stand on its own.[2]

Another difference between the Gandhian and Nehruvian visions of development was Nehru's emphasis on industrial development and top down national economic planning, in contrast to Gandhi's vision of an agrarian society in which development would happen from the bottom up. Chandhoke (2005) argues that Nehru's vision of a state-driven, planned economy was influenced by experiences in several parts of the world where this model of development had lead to successes. These included the Soviet Union, the US and its Keynesian policies of the 1940s (which were a response to the economic depression of the 1930s), and ironically, developments in the country of the former coloniser, England, where a welfare state had emerged from the struggles of the working class in the first half of the 20th century. According to Chandhoke, even important actors in the Indian commercial sector and the influential industrialists of the 1940s and 1950s agreed that the government should have a central role in planning and structuring the economy. There was broad consensus that state control and the extension of development benefits to everyone 'were deeply interrelated (2005:1038).'

The politics of top-down economic planning and control continued into and through the period of Indira Gandhi's leadership in the 1970s.[3] In the 1980s a serious debate emerged as to whether the planned economy should be abandoned and India's markets opened to the global economy. Rajiv Gandhi, India's prime minister in the late 1980s was the first Indian leader to envision consumption and the middle class as potential engines of economic development (Fernandes 2000). This was a radical change from the visions of both Nehru and Gandhi. As Fernandes wrote, 'If the tenets of Nehruvian development could be captured by symbols of dams and mass based factories, the markers of Gandhi's (Rajiv) India shifted to the possibility of commodities that would tap into the tastes and consumption practices of the urban middle classes (2000:613).'

This change in the place of consumption in Indian development was made during a period of declining economic growth, growing national debt and a growing trade deficit (Khilnani 1998; Corbridge and Harriss 2000). The Gulf War of 1990 exacerbated these problems because, as explored in Chapter 6, it led to declining revenues from Non-Resident Indians. It also led to higher oil prices, which put stress on the Indian economy. The collapse of the Soviet Union in the late 1980s contributed to a decline in foreign trade and 'removed the only alternative ideological model to the capitalist market (Khilnani 1998:96)'.

These developments strengthened the hand of the proponents of liberalisation and the 'opening' of India's economy.[4] A Congress-led government was elected in 1990 and its Minister of Finance, Manmohan Singh, announced a package of liberalising reforms.[5] Conditions would be relaxed for foreign imports, investments and joint ventures between Indian and foreign-based companies. The government relaxed import restrictions and laws on foreign investment. Further, it took an International Monetary Fund (IMF) loan of USD 2.8 billion, bringing with it demands for deep structural reforms to the Indian economy, including downsizing the public sector, privatising public enterprises and establishing conditions for freer markets (Corbridge and Harriss 2000). TNCs, which had been virtually locked out (or thrown out as in the case of Coca-Cola's expulsion in 1977) were invited in. The last vestiges of *swadeshi* were swept aside. When journalist Jeremy Seabrook (1992:9) questioned one of the reform-friendly members of the Indian Planning Commission about possible negative consequences of the 'opening' of India, his answer was 'You're not going to give me any of this Gandhian bullshit, are you?'

The 'opening' of India legitimised and encouraged consumption as an integral part of both national and personal development. Indian sociologist Rowchowdry wrote that after 1991 'the awesome power of a predominantly consumption-oriented economic model worked at multiple levels to create a reconfiguration of popular images of what is a desirable life' (2001:10). At the individual level, consumption, or the pursuit of a consumer-oriented lifestyle, took its place alongside Indian identities such as 'moral agent, citizen, associate' (Lury and Warde 1997:101).[6] This national and personal legitimisation of consumption was accompanied by a massive change in the range of new products available and affordable to middle class households after 1991. The policy of removing import barriers, allowing direct investment and collaborations between foreign and Indian companies, and the lowering of import duties and luxury taxes all contributed to a rapid movement into India of not only new products but completely new product categories (for example in the domains of cosmetics, cleaning products and refrigeration). For the first time, foreign companies were allowed to have majority shares in Indian corporations. TNCs from North America, Europe, Japan and Korea used direct investment and subsidiary creation to access new markets, accompanying these with new financing schemes and heavy marketing. According to Lall, the number of foreign collaborations between transnational and Indian companies grew rapidly from 950 in 1991 to 2303 in 1996. During the same period the amount of foreign investment grew from 5.3 billion rupees to 361.5 billion rupees (2000:200).

Within a few years after 1991, TNCs or their subsidiaries had made inroads into the production and sales of household appliances, cars, cosmetic and soap products in India. To name a few examples, Whirlpool (USA) and Electrolux (USA) bought existing manufacturing facilities (Maharaja International and Kelvinator, respectively). Other TNCs established subsidiaries or expanded their previously established subsidiary operations. Some formed joint ventures or 'strategic alliances' in order to take advantage of established local distribution networks. Sanyo (Japan), Matsushita (Japan), Daewoo (South Korea), LG (South Korea), Maytag (USA), Whirlpool (USA) and General Electric (USA) are some of the TNCs that formed joint ventures with Indian companies or established subsidiaries in the 1990s.

In the emerging Indian political discourse on TNCs and global markets, the Congress party was a proponent of liberalisation and openness. The Hindu nationalist parties such as the Bharatiya Janata Party (BJP) were divided in their views. One segment of the nationalists revived the Gandhian vision that equated true Indian development with the barring of Western influence and products. Others 'saw a willingness to leap into global markets as the distinguishing mark of a self-confident nation (Mazzarella 2003:9)'. For proponents of the 'leap', a cringing and protectionist attitude was seen as fostering a perception of India as a second rate player on the international scene. Yet others took the pragmatic view that excluding TNCs would hamper India's efforts to develop its own high-tech industry. The convergence of a consensus among the neo-liberal, pro-Western nationalists and political pragmatics contributed to a national political stance that broadly supported the 'opening' of India. Varma contends that this political turning point marked 'the collective exorcism from the nation's psyche of the repressive and life-denying nature of Gandhi's idealism'. He writes that after 1991 'material wants were suddenly severed from any notion of guilt' (1998:175).

In spite of Gandhi's 'exorcism' from the political debate about foreign products and consumption, the anticipated frenzy of consumption of foreign products did not materialise right away. The initial wave of TNCs to take advantage of the liberalised conditions in India did not meet with overwhelming success. Just because a product was popular in other parts of the world did not guarantee its success in India. Experiences in the first half of the decade of the 1990s showed that TNCs and international advertising agencies would have to adapt their products to local tastes and adjust their marketing messages if they were to be successful (Mazzarella 2003). For some, this initial stumbling of multinational companies was a source of pride. As one political columnist cited by Mazzarella put it, 'Successive waves of transnationals have been unable to colonize the country's customers'; another wrote that the failures of TNCs compensated for 'the humiliation of the colonial experience' (2003:263).

In Kerala, consumers did not uncritically rush out to buy products just because they were made by a foreign manufacturer or had a foreign label. However, there is no doubt that the stigma that Gandhi had associated with foreign products had dissolved by the end of the 1990s. An elderly Hindu Ezhava man told me: 'Before it was important (to buy Indian goods). Today the Gandhian ideal of "buy Indian" is dead.' When I asked Preethi and Vibek, a young Hindu Nair engaged couple, whether Gandhi's ideas on Indianness and frugality had relevance for their generation, they responded simultaneously with an emphatic 'No!' When I then inquired 'Maybe for your father's generation?' Preethi responded with a touch of irony, 'Grandfather's generation maybe'. Sindhu (Hindu Nair in her 20s) put it this way: 'We have a great feeling (for Gandhi) but I don't think we practice his ideas. In our family we like things from other places. We buy things that come from outside also.'

Many people equated their greater access to foreign goods with freedom from the repressions of colonialism. A middle-aged Hindu Ezhava man told me:

> Each and every day there are new developments. We also have to participate in that. In this period we can't say that it is a foreign idea and avoid those. Before independence English people did not give us anything. They were only taking from us. They took from here and they produced the output there and sold to us. To get knowledge

about the world and to go along with the movements of the world this is necessary. This was a part of the freedom movement.

Liechty found a similar association of freedom with consumption among the middle class in Nepal: 'In Kathmandu, the classic middle-class construal of "freedom" as "freedom to consume" takes on a different meaning; modern consumer practice has to be seen in light of a past in which even those with resources were simply not "free" to consume (2003:125).' In Kerala, however, most people were less interested in this political or symbolic side of consumption and more interested in the pragmatic advantages of having access to better quality products and lower prices. A 39-year old Hindu Ezhava man expressed it this way:

> The common people like us want to get quality goods at low price. Even if it is imported we don't care about it. The common people have to think only about the price and quality. If there is a competition in the market then the price will come down. If there is only Indian products then the price will be high. That is why the Chinese goods are allowed to be sold here. We are only thinking about the quality and price of the goods. We don't have special affection to the foreign things or love Indian products more. If I buy clothes or cassettes I buy things from here. I don't have a craze for foreign goods.

Gopal, a Hindu Ezhava who had grown up in Kuwait echoed this point: 'It (opening to foreign products) has improved the quality of the products. For example these things we buy over here like powder and soaps and my brother when he comes from the Gulf, he sees a great change in the quality of products.'

There seems to be no doubt that the Gandhian ideals of 'buy Indian' and live frugally are no longer relevant to the ways people consume. When I asked one 40-year old Hindu Ezhava man what he thought about 'buying Indian', he replied, 'Gandhi said that a long time ago...If he said that today they would tell him to get out. Everyone is interested in imported things. They are good quality and low price.' My Muslim friend Anil said that today 'Even the oldest of old men will use the foreign products.' An elderly Hindu Brahmin woman told me:

> I think it (the availability of foreign products) is quite good! We know (then) what other people (outside India) are using. And better appliances, it is better to go for that. And so many changes are coming. It is nice. I feel we have to move along with it. In my opinion we have to

preserve our Indian culture, but if we use all these appliances that is not affected.

Her idea that Indian culture would not be affected by foreign products is something I heard often. Kerala could 'move along' through consumption, and could do so without destroying its 'culture'. When I asked her what she meant by culture, her response was also typical of what I heard from many others: 'We have our own religion, feasts, our way of clothing – like wearing bangles.' Neither she nor any of our other neighbours in Kumarapuram seemed to be worried about any potential negative consequences of global markets or global media for Kerala's cultural traditions.

To summarise thus far, Gandhi contributed to a political and personal view of consumption that linked Indian identity and its preservation to material simplicity and insulation from the materialist West. Nehru was less concerned about total insulation, but was wary of developing an economic dependency on the West. There was a gradual erosion of the politics of economic insulation over the decade of the 1980s and a significant relaxation of restrictions on foreign investments and products after 1991. This by no means resulted in a realignment of Indian consumption tastes to mimic those of the West. In fact TNCs experienced setbacks in the early 1990s and in order to increase their sales had to reorient product designs and the ways products were marketed to take account of Indian cultural differences. The concept of 'close distance', proposed by Mazzarella (2003), captures very well what Malayalees want in their products: a sense of participation in global developments without surrendering Indianness. I return to this point in the discussion of TNC advertising in Chapter 9.

## Kerala socialism

As an Indian state, Kerala was 'opened' to global capitalism in conjunction with the opening of India as a whole. However, Kerala differs from other states (with the exception of Bengal) in its history of left-leaning, socialist lead governments. As outlined in Chapter 2, coalitions of left and left-centre governments have been in power in Kerala throughout much of its history. This has lead to political goals aimed at redistribution of wealth, a planned economy, and significant public spending on health and education. The approach has been dubbed the 'Kerala Model' (see Dreze and Sen 1996) and has been widely studied and promoted as an alternative to neo-liberal economic development models.

In this section, Kerala socialism is explored, giving attention to middle class participation and why India's highest consuming state after 1991 is one of the two states with a history of socialist governance.

Surendran (1999:30) summarises the basic elements of the Kerala Model this way:

> The Kerala model of development may be defined as a long term strategy which enables a country or a state or a region to achieve a higher physical quality of life even at a low growth in productive sectors and a low domestic income by diverting the major share of its available resources for the creation of infrastructure for human resource development like education, medical care, housing and sanitation, etc. The conceptual definition of Kerala model of development highlights four tenets:
>
> 1. It stipulates that economic growth is not a pre-condition for the attainment of economic development...
> 2. Socio-economic awareness of the people may be a pre-condition of economic awareness. The cultural upward mobility in a region may further lead to economic upward mobility.
> 3. The transfer of land resources from landlords to tenants or agricultural labourers may be a pre-condition for sustaining social justice which, in turn, leads to economic justice and prosperity.
> 4. Rapid investment in social infrastructure like education, health care, housing, road and communication along with productive infrastructure is a pre-condition for development.

As Parayil (2000:12) put it, the Kerala Model was founded on a belief that 'poor societies are able to improve the conditions of living of their people without waiting first for economic growth to materialise'. The first government of the newly constituted State of Kerala in 1958 gave priority to 'human development', which has been pursued by most Kerala governments since, whether left or right. The results of the Kerala Model have been impressive, measured in terms of health and educational achievements. As formulated by Issac (2000:4), 'Kerala's development ... shows that even at low levels of economic development, basic needs can be met through appropriated redistribution strategies...'.

The precepts of redistribution and public ownership of essential services (health, energy, water) have characterised the programmes of Kerala governments of both left and right. Kerala has invested more heavily in health and education than any other Indian State. Working conditions and workers rights are well-defined and protected. Minimum wage and average salaries for government employees are among the highest in India. Kerala has India's most literate population. The Kerala Model and its achievements were an important source for the United Nation's Human Development Index, which were based on achievements in literacy, health, education, life expectancies and other indicators of well-being.[7]

In the 1990s, an ambitious project was initiated by the government, aimed at extending the benefits of the Kerala model to previously excluded groups.[8] The 'People's Plan' aimed at encouraging decentralised, participatory democracy and the fostering of local economic development (Isaac 2000). The Plan drew on work of the *Kerala Sasatra Sahita Parishat* (KSSP, translated People's Science Movement). Its members were mainly middle class academics and scientists. Beginning in the 1970s the KSSP had carried out projects with local people, intended to retain, or in some cases to recapture local knowledge about management of agriculture, forest management and natural resources.[9]

The work of the KSSP encouraged local participation in governance and economic decisions. The People's Plan made use of this experience. According to Tornquist (1995), some 10 per cent of the entire population of Kerala (three million people) took part in the participatory process of developing local plans for self-governance and local development. New political forums called *Gram Sabhas* were created that represented a broad range of local people and their interests. Plans were discussed, approved and sent via the local districts to the State Planning Board. Public funding was distributed for implementation of the Plan by local authorities. The first plans were implemented in 1998.

In addition to political inclusion, the Plan also aimed at establishing local economic entrepreneurship through the setting up of 'self-help' groups in rural areas. These were voluntary groups of families and individuals who could apply for small grants to supplement their own common pool of resources, usually accumulated through a small monthly contribution by each participating family. In some communities, the collective holdings were formalised in the form of a 'people's bank,' which offered zero or low-interest loans for local business initiatives. One people's bank in Kunnathuka, a district located just south of Trivandrum loans labour rather than money. Local participants form a labour pool from which farmers can borrow labour in peak periods such as harvests.

Self-help groups have been active in establishing local enterprises through alliances with locally-funded government or NGO programmes;

for example, the programmes of the State Government's Energy Management Centre, which trains local women to produce and sell an energy-efficient thermal cooking device (*thaapabharani*) (Wilhite 2002). Another example is an alliance with the *Swadeshi* Programme, run by the Centre for Gandhian Studies at the University of Kerala. Training programmes run by the Centre teach participants how to set up small-scale, local production of mainly soaps and food products, using locally available ingredients. In 2002, training programmes had been conducted in 300 rural centres around Kerala. The local production of soaps and foods employed 15,000 people in 2002.

What we can derive from Kerala's democratic socialist governance is that the political contexts framing many aspects of everyday life in Kerala have emphasised social levelling, collectivism, and public (rather than market) mechanisms for steering economic activity. While there has not been an active political discourse discouraging consumption or encouraging thrift in Kerala, as was the case in post-World War II Japanese socialism (Garon 2003), neither have the political contexts around consumption actively encouraged private consumption. Another important point is that Kerala's socialist projects involved the participation of many Malayalee from all walks of life, including the middle class. Franke and Chasin's analysis of Kerala socialism concluded that 'Kerala's activists have shown an ability to mobilize very large numbers of people for a variety of causes', pointing to peasant movements, trade unions and literacy campaigns from the 1950s through the early 1990s (1996:7). Robin Jeffrey (1992) analyses early social activism in Kerala and writes that it has been noteworthy because of the involvement of middle class and elites in class and caste activism. The middle class contributed to the fundament of the Kerala Model and through land reform, educational reforms, caste reform and redistributive policies, the Kerala Model has in turn contributed to the growth of the middle class.

In the Kerala state elections of 2004, a coalition of left-leaning parties was returned to political power after a brief period of control by a centre-right government. The voter turnout was very high, in the range of 70 to 80 per cent, an indication that the vast majority in Kerala, including the middle class undoubtedly still largely supports the precepts of the Kerala Model. At the same time, however, consumption of public services is declining. Many new private hospitals and clinics have sprung up, catering to the growing middle class. Concerning education, in 2003, more than half (56 per cent) of the pupils in lower grade schools from one to ten were enrolled in private schools. A Syrian Christian couple, Jacob and Meena are strong advocates of public education. Their children are enrolled in a public school in a village outside Trivandrum. However, they told me that the teacher-student ratio is much lower in their children's public school than in private schools in the area, and that limited budgets meant that students had limited access to new books and computers. These issues were a source of consternation for Jacob and Meena, fearing that they were sacrificing their children's educational future for their principles. They were strongly considering moving their children to a private school.

Given the increases in private consumption in domains that have previously been dominated by consumption of public goods and services, taken together with the increasing importance of new forms for private consumption, one can deduce that the majority of the Kerala middle class see no contradiction between a leftist political orientation and a high consumption lifestyle. R. Parvathi Devi, a former member of the CPI-M and a popular television journalist told me that 'People are taking as enthusiastically to consumerism as they did to communism. People crave consumer goods. They rate their success in life in terms of goods. Children of both the left and right are interested in consumption.' Based on my study, the view that people 'crave' consumer goods is an overstatement. Nonetheless, one can say that socialism as practiced in Kerala has not provided any significant friction to the growing interest and participation in consumption.

## Religiosity

Another apparent anomaly in Kerala is the growing importance of both the pursuit of new goods *and* the pursuit of religion. Interest in religious participation spans peoples of all religions and Hindu castes. In previous chapters, we have seen how religious practice is an integral part of everyday routines. Approximately two-thirds of the survey participants of all religions in Trivandrum said that they participated in religious activities outside the home at least once a week. For Hindus, this is because religion is so thoroughly integrated into home practices and everyday lives. The Hindu scholar Varma (1998) asserts that Hinduism is as much a way of life as it is an organised religion. 'Hinduism...has no organised Church, no one God, no paramount religious text, no codified moral laws and no single manual of prescribed ritual (1998:123).' In the absence of a strong centralised church, local temples and homes are important sites of religious practice. Almost every Hindu home in Trivandrum has a prayer room or prayer corner. Prayers are made daily and food is regularly offered to ancestors and gods. Outside the home, many Hindus regularly participate in temple celebrations, processions and pilgrimages. Religious festivals permeate the calendar.

Religion in Kerala is not just something for the older generation. In fact, many older people claimed that their children were more interested in and knowledgeable about religion than they were. One of our elderly Hindu Nair neighbours told me that all of his four children, their spouses and his grandchildren were more knowledgeable about Hinduism than he and his wife were. He said of his oldest son, 'He follows all the traditions. He gives me lectures. He has read all the Vedas and everything.'[10]

## Pilgrimages

Men and women of all ages participate in religious pilgrimages and retreats. The Hindu pilgrimages such as *Sabarimala* (mainly intended for men) and *Attukal Pongala* (mainly for women) were introduced in Chapter 5. The interest in *Sabarimala* pilgrimage has been growing over the years, attested to by the increasing number of participants and by the huge viewership of live television productions from *Sabarimala* (Younger 2002). Despite the rigours of the journey, in 2002 over ten million people made the pilgrimage to visit the temple during the period that the idol of the God *Ayyappan* is on display (end of November until end of December), or about 20,000 visitors per day. In the middle class neighbourhoods surveyed in Trivandrum, half of the male heads of household had attended the *Sabarimala* pilgrimage at least once in the previous three years.[11] The pilgrimage involves a significant commitment of time and effort. First, are the preparations, involving fasting and not shaving for a period of weeks; second is the journey itself, which involves an overnight car or bus trip from Trivandrum and a strenuous walk up the mountain to the temple. The walk is not a Sunday stroll. It is made barefoot over a rocky path in the pre-dawn hours. Pilgrims bear a heavy bag on their heads filled with coconuts and other gifts for the God *Ayyappan*.[12] The televised programme broadcasting from *Sabarimala* has one of the highest viewing audiences of all Kerala programming. At *Makadam* (in January), legend has it that a light flashes behind the mountain, visible only to those pilgrims who are pure of mind. The television programme brings the hope of seeing the light to non-pilgrims at home in their own living rooms.

## Devotionalism

Interest in another form for Hindu religiosity, devotionalism is also increasing in Kerala. Devotionalism has a long tradition in Hinduism.

In the past, devotionalism was mainly confined to members of the Brahmin caste, but today it is widespread among the Nair and Ezhava castes. Fuller (1992) found this also to be the case among the urban, middle class in Tamil Nadu. Many of the Hindus in Trivandrum middle class neighbourhoods devote their prayers and financial support to a guru. The gurus with the largest followings in Kerala are Sathya Sai Baba and Sathguru Shri Matha Amrityanandamayi, also known to her followers as Amma (mother in Malayalam). Both see themselves as avatars, representing deities on earth. Amma associates herself with the god Vishnu, having taken the most popular Vishnu motif and replaced his face with hers on posters tacked up all around Trivandrum. She has recently added Devi to her name, associating herself with the Hindu goddess Kali. Amma's principle Ashram is in Coilan, 50 kilometres north of Trivandrum. Her Kerala-based movement has gotten considerable international attention. In 1993, she was invited to speak at the 50th anniversary celebrations of the United Nations and in 2000 she was invited to attend the World Peace Summit in New York. Many Trivandrum middle class Hindu men and women are Amma devotees. Most of them regularly donate money to her cause. When I asked what they expected in return, a typical response was 'good fortune' (as one of our Nair neighbours expressed it).

Sathya Sai Baba's main ashram is at Puttapurthi, but he also maintains colonies and residences in Karnataka, Andra Pradesh, Mumbai and Chennai. Sai Baba is the self-proclaimed reincarnation of a Holy Man also called Sai Baba. He has devotees among middle class and elites all over India. In an analysis of the reasons for his popularity, Babb (1986:169) suggests that it is due to two things: first, that guru worship represents a way to re-enchant secularised and material lives; and second, that this form for religiosity is a form for what he calls light Hinduism. 'The cult of Sathya Sai Baba does not make great demands on its adherents...the imposition of rigid rules is foreign to the spirit of Sathya Sai Baba's teachings.' One can reaffirm one's Hinduness through devotion to Baba without renouncing modern lifestyles.

Baba does not take an explicit position on the pursuit of wealth or material goods, but it is clear that he does not discourage or disparage high consumption lifestyles. Our Hindu Nair neighbour Meena told me that Sai Baba offers her 'peace of mind' in return for love, tolerance and service to the poor. Hindu Nair Gopal, a follower of Sai Baba, told me, 'Baba is supposed to help rich people to help others.' Gopal believed that by dedicating himself to Baba, a dedication that included regular donations (one can donate to Baba's Trust at any branch of the Canara Bank, which has branches nationwide, including Trivandrum), he would be rewarded by good fortune. As Babb (1986:160) writes, Sai Baba represents a kind of 'modernized saintliness' that corresponds very well to the needs of middle class Indians, who on the one hand yearn to be cosmopolitan and wealthy, and on the other to stay in touch with 'the ancient' worship of gurus.

A new popular form for devotionalism is the 'New Age' devotional movements. The 'Art of Living' is the most popular of these in Kerala. These movements are interesting cases for globalisation studies. The inspiration for them originated in India, the idea was then exported to the West and taken up in the New Age movement, after which it made its way back to India. The first of these global 'u-turns' was initiated in the 1960s by Maharishi Mahesh Yogi. His 'transcendental meditation' gained a huge following in the USA and has since spread to many other parts of the world. This and several other devotional new age movements returned to India in the 1990s.[13]

Movements such as the 'Art of Living' emphasise prayer, meditation and the realisation of 'peace of mind' for the individual participants. 'Art of Living' is centred on the guru Shri Ravi Shankar. It offers tuition-based courses on 'self-realization' through the mastery of meditation achieved through a breathing technique called *pranapara*. Ravi Shankar is a popular speaker and a frequent television talk show participant. Many of our neighbours had attended at least one 'Art of Living' course. They insisted that my wife and I attend the beginner's course. Based on that experience, as well as reading 'Art of Living' literature (The Art of Living 2003) and the speeches of Ravi Shankar, my impression is that 'peace of mind' can have elements of both the spiritual and the material. As with the more established devotional movements, the pursuit of material goods does not interfere with the attainment of spirituality.

Many of the Hindu spiritual leaders point to material simplicity and renunciation as important aspects of Indian philosophy and as relevant to understanding the way people consume. For example, Hindu spiritual leader Kulkarni (2001), wrote in a regular column in *The Hindu* newspaper on religion:

> The Western Philosophical tradition has a mechanical world view and materialistic way of life, while the Indian philosophy has culminated in an organic, holistic and spiritual world view and renunciative way of life. Indian philosophy places before us the goal of

eternal bliss which can only be derived by reduction in wants and hence people at large opt for a renunciative way of life and their consumption reduces.

Looking closely at the exercise of the various forms of Hindu practice, Kulkarni's points do not seem to have much relevance. The promotion of thrift, simplicity or frugality is not an important element in either conventional Hinduism, pilgrimages or devotionalism. Fuller (1992) supports this point in his review of Hindu doctrine and practice. He points out that there is a Hindu Goddess responsible for material and commercial prosperity (Lakshmi). Based on my observations and interviews, no one of any religion or caste equated the achievement of spirituality with a 'reduction in wants' and a 'renunciative way of life'. Virtually no one had reflected on the possible contradictions between exercising their religion and exercising their interest in acquiring new things. Waldrop's (2001:219) conclusion about religiosity among elite families in Delhi resonates with this: 'People I interacted with are thus religious, but their religion seems to be rather goal oriented. They pray to get a daughter into Cambridge or fast in order to get a daughter married.' Osella and Osella (2003:747) observed that Sabarimala participants 'go hoping to get jobs, improve their prospects, or keep their businesses running smoothly.'

### Christian religiosity

Christians of all of the many branches and denominations found in Kerala are active in the exercising of their religion. Churches are important loci for events virtually every day of the week, and Christian pilgrimages and retreats are popular. An important pilgrimage for Catholics involves a visit to the shrines of Kerala Christian saints, located mainly in central Kerala around Kottayam (Dempsey 2001). Another involves participation in group retreats such as the popular Mulinoor retreat. The Mulinoor retreat lasts for two weeks and is attended annually by up to 20,000 people. In more than a third of the Christian families surveyed in Trivandrum, either the husband or wife (or both) had attended the Mulinoor retreat at least once in the previous three years. Our neighbour Sarida (Syrian Christian) had attended the retreat three times. She was enthusiastic about the experience and emphasised its spirituality. 'From morning until late night there are prayers. And music. Very spiritual and very nice. A very good experience. Something spiritually very good.'

As in Hinduism, Christian doctrine has contradictions and inconsistencies concerning the pursuit of material goods. Miller (2001a) examined Christianity in Europe and North America and found examples of how both moderation and consumption have been encouraged. He cites the practices of feasting and fasting, as well as the combination of materialism and aestheticism in traditional religion under Calvinism (referring to Schama 1987). Christmas is an example of a Christian celebration involving both consumption and spiritual reaffirmation. Wilber (cited in Nash 1998) points to contradictions on the subject of materialism in the Bible and to a long debate in Western Christianity on the morality or immorality of the pursuit of material wealth.

In Kerala, Dempsey (2001) found that the dedicated Catholics of central Kerala have strongly held ideas about the sinfulness of avarice. They see their religiosity as an 'antidote to material excessiveness... (and) the West itself (2001:27)'.[14] In Trivandrum, I did not find any Christians of any of Kerala's denominations – Protestant or Catholic, Syrian or Roman – who saw a contradiction between their spirituality and their consumption of products, whatever their origin. When I asked Sarida about the relationship between spirituality and consumption, she responded, 'We have to remember God in everything we do. But we are not denying the material life.'

I discussed religiosity and consumption with Rija, our insightful young (Hindu Ezhava) neighbour. She suggested that many people engage in religion as a strategy for coping with the complications of modern lifestyles, or as she put it, 'difficulties in dealing with more luxurious lives.' At first, I thought she meant that people engage in religious activity to compensate for a bad conscience about their interest in material goods. After further reflection, I realised that what she really meant was that people see religiosity as a source of help in dealing with the problems and complications brought on by 'more luxurious lives'. She gave examples of the kinds of problems and conflicts she meant: disagreements within the family over new purchases; the strains of incurring debt or buying on credit; and the strains on household budgets due to the costs of fuel for cars and electricity for new appliances. Further, there are the strains of keeping up with, or staying abreast of neighbours. Liechty found something similar in his study of middle class consumption in Kathmandu (2003:34): 'The pleasures of consumption, while real, are never far removed from the gut-wrenching anxieties that arise as people attempt to maintain their positions in the middle-class consumer culture.' Syrian Christian Sarida told me, 'Nowadays tension and all of these things are much more than before. Even children are facing tension. People are facing so much problems nowadays. So people are praying to God. And they are going to these retreats and all these things.'

## The decline of frugality

Neither consumption nor foreign goods are today seen as antithetical to middle class Indian identity. For middle class consumers in Trivandrum, 'buy Indian' is a slogan that no longer has relevance for their consumption. As we have seen, many regard the consumption of foreign goods as an expression of freedom from global repression. People want to be part of developments in the wider world and to have access to things that they feel will improve their lives. One can be a committed socialist without denouncing consumption. In an ironic twist, socialist projects may have increased the consumption potential in Kerala by enlarging the middle class (through land reform and affirmative action programmes) and creating a fully literate society.[15] According to Consumer's International (CI-ROAP 1998), Kerala's well-educated population makes it a favourite place for the testing of new products and new marketing campaigns.

The frugality and thrift that were noteworthy characteristics of the Indian and Kerala middle classes of much of the post-Independence era are dissipating as desired social values (Gupta 2000). High savings rates could be seen as an indicator of frugality. Garon (2003) analyses Japanese consumption, using the relationship between household savings and spendings an indication of frugality. General studies of both Indian and Kerala household savings rates show that they have increased in tack with increasing incomes over the past decade. A recent Government of India (2005) report shows that savings rates are increasing, but at the same time increasing consumer expenditures are the major contributors to economic growth since 2000. I would argue that in periods of economic growth, high savings rates are not a good indicator of frugality. Garon would seem to support this when he writes about Japan after the 1980s: 'Among the nation's households, the inculcated habits of thrift (measured in terms of savings) did not diminish noticeably but rather coexisted with the new consumption (2005:205).'

Many people in Trivandrum talked about the changes in attitudes to frugality between their grandfathers' generation and their own generation. Syrian Christian Abraham said that for his parents' and grandparents' generations, 'Even if you were rich, you did not show it.' Nash (1998) found this same dissipation of the importance of frugality in Europe: 'Frugality has certainly not been fully forgotten, but it has been demoted ... Frugality today is often greeted with amusement, ridicule, or even contempt (1998:417).' There is evidence for the contempt for frugal lifestyles in Kerala. A frugal lifestyle is regarded at best as old-fashioned, and at worst as an indicator of poverty.[16] Hindu Ezhava Sateesh talked about the many ways that his grandfather exemplified the value of frugality and simplicity through his lifestyle. His moderation in all things was a source of great respect in his community. Sateesh had modelled himself after his grandfather. However, Sateesh felt that his own frugal lifestyle had brought him and his family scorn rather than respect. An example was his principled decision to not connect to cable television because of the access it would bring to Cartoon Network and its negative influence on his children (especially its heavy advertising). When his children would visit neighbours houses to watch television, his neighbours began gossiping about what they characterised as miserly behaviour. He eventually succumbed to social pressure and subscribed to the cable network.

Frugality was actively promoted in Gandhian *swadeshi* and there is evidence that it was seen as an important trait in middle class consumption. Its importance is dissipating as an agent in constraining Kerala consumption. The next chapter examines the ways in which television programming and advertising are implicated in the decline of the importance of frugality in Kerala.

# 9
# Television: 'Everyone is Watching'

Virtually every middle class Trivandrum home is equipped with three things: a mixmaster (mixie), a designated place for prayer and a television. This says much about the importance for Malayalee of food, religion and television viewing. The first two of these (food and religion) and their relation to consumption has been explored in previous chapters. The relationship between television viewing and consumption is the subject of this chapter.

When I asked Hindu Ezhava Ajith (in his 60s) what had changed most about social life in Kerala over the past 20 years, he pointed to his television and said *'Everyone* is watching.' Almost everyone interviewed from middle age and upwards echoed this point. Television viewing has replaced cinema as the most important source of entertainment in Kerala. Its popularity was vastly increased after the mid-1990s, when cable television became available. In the 1980s, access was limited to only three Indian channels. From the mid-1990s Trivandrum residents have had access to scores of cable channels broadcasting dramas, variety programmes, news broadcasts, talk shows, movies and sports from Kerala, other parts of India and abroad. At almost any time during the day in the homes of our neighbours in Kumarapuram, if someone was at home, the television was on. Television has had the greatest impact on the daily lives of women, many of whom are at home during the day and for whom the television is an important source of entertainment and information. The most popular programmes are the Malayalee evening serials (soap operas). After 1930, many of our Kumarapuram neighbours settled in front of the television to watch the serials on the local channels, Asianet or Surya. The most dedicated viewers are women. While men usually feigned a lack of interest, I found many who were surprisingly up to date on plot developments.

Television viewing and role in everyday practices have not been important anthropological topics (with a few exceptions such as the contributions to *Media Worlds* by Ginsburg, Abu-Lughod and Larkin 2002, appropriately subtitled *Breaking New Terrain*). One reason for its absence in anthropology is that the television has not been present in the homes of typical anthropological subjects until recent decades. However, one can no longer argue that television is peripheral to everyday life in many of the regions patrolled by anthropology, including India. Of the households surveyed in Trivandrum middle class neighbourhoods, 96 per cent had a television. Women watch television on average four hours daily. In the evenings, popular programmes draw people off the streets and thereby affect the fabric of social interaction.

Television programming and advertising have become unavoidable subjects for consumption studies in places like India. However, both the theories of television's effects on behaviour and methods for inquiring about the relationship are 'contested and fragmented' (Spitulnik 1993: 293). The early theoretical approaches originated from cultural and media studies. The research pioneers of television in the 1950s and 1960s mainly subscribed to the theory that the television and its programmes manipulated passive audiences. The research question was how passive audiences reacted to programming (Packard 1957; Hall 1973). Marcuse (1964) was one of the early theorists who argued that television created false needs and encouraged social conformity. From the 1970s, the theory of manipulation evolved into an acknowledgement that audiences are reflexive and intelligent. Interpretations and reactions to programming were not uniform, and effects on practice subtle and varied (Story 1999; Morley 1995).

As televisions moved into homes in Asia, Latin America and Africa, the early theories about passive audiences moved with it. The television was viewed in early work as a kind of Trojan horse, bringing 'the other' and ideas about it into homes in the South and changing local ideas about 'the good life' (see Featherstone 1990 and Tomlinson 1991). This theory does not work in India today, first because audiences are active (critical and reflexive) but also because the 'other' that they are exposed to is far from uniform. Even assuming that television programming was manipulating, programmes from North America, the Arabian Gulf, Russia, Japan and elsewhere bring vastly different ideas about 'the good life' into India. Bringing these 'others' into Kerala living rooms undoubtedly makes people reflect about other ways of doing things, but certainly does not have a homogenising effect on practices.

In Kerala, locally produced programmes are vastly more popular than foreign programming. As will be developed below, these programmes both reflect and reproduce local socio-cultural practices and norms. The popular evening serials draw people to the television night after night and expose them to massive amounts of advertising. From a consumption perspective, this complementary relationship between popular programming and television advertising is important. Before exploring the local programmes and their advertising, I briefly outline the history of television in India and Kerala from its inception in the 1950s.

## The advent of television in India and Kerala

In her book *Screening Culture, Viewing Politics,* Mankekar (1999) outlines the history of Indian television. Early television was totally owned and controlled by the Indian government. The government established a national network called Doordarshan in the 1950s, which had control over eight television stations. The government saw television as a means to contribute to nation building and to consolidate a sense of collective Indian identity. In the 1970s Doordarshan began to experiment with new types of programmes. These included situation comedies, dramas, musical programmes, quiz shows, and, the 'serial', a story broken up and shown in many regular segments. Doordarshan purchased the rights to popular foreign serials such as *Dallas* and *Falcon Crest,* but Mankekar writes that after an initial burst of interest, people did not watch them in great numbers. The first serial produced in India was called *Hum Log,* which was immensely popular. Its 156 episodes garnered a huge television audience. Other Indian serials followed in its wake.

In the early years, the government had imposed strict regulations on the time that could be devoted to advertising. Producers were not allowed to interrupt programmes with advertisements. They were required to be grouped together and shown between programmes. Early on, the civil servants in charge of Doordarshan saw that the serials provided a huge potential for commercial advertising. By 1975, television advertising revenues only constituted 1 per cent of Doordarshan's budget. Over the next 15 years, restrictions on advertising were relaxed and advertising revenues grew. In 1990, advertising contributed 70 per cent of Doordarshan's budget, the equivalent of USD 300 million. Commercial sponsors were eager to sponsor the Indian serials because of early surveys which showed significant effects on sales. Mankekar points out that the principle sponsor of *Hum Log*, Maggies Noodles, increased its sales by 300 per cent in the two years after it began sponsoring *Hum Log* in 1982.

Indian television changed dramatically in the decade following 1991. Rules on foreign investments in Indian broadcasting were relaxed to allow foreign networks to establish joint ventures with Indian companies. A few years later, broadcasting was fully globalised, allowing foreign-based networks to broadcast in India. CNN and StarTV (based in Hong Kong) began broadcasting in India in the early 1990s. StarTV brought movies, dramas and entertainment programmes from all over the world, including the popular international beauty pageants discussed in Chapter 3. Eurosport brought sports such as rugby and European football. Cartoon Network, based in the USA, brought 24 hours daily of programmes and advertisements aimed at children. CNN, BBC World, French TV5 and Al Jazeera brought news from almost every part of the world. These global news stations bring Malayalee in touch with real time viewing of world events, shaped by many of the same news agencies that shape the news in the United States, United Kingdom and the Middle East.[1]

Foreign and local programmes of all kinds are popular, but by far the most popular programmes are the Hindi and Malayalee serials. As indicated above, the serials are an Indianised form of the 'soap opera', which took form in American radio programming of the 1930s. Soap operas were purposefully designed to be never-ending. As Gledhill writes, the soap opera broke with the idea that a story should have 'a beginning, a middle and an end (1997:368)'. It drew its popularity from a cycle of 'equilibrium, disruption, equilibrium restored in the lives of its characters'. The soap opera made its way into television programming in the USA in the 1950s. The generic television soap opera plot centred on the lives of women and on 'women's concerns' (Barker 2000:267). 'Women were faced with dilemmas, or oppositions, which were played out through an unending serialisation (Gledhill 1997:366).' This describes perfectly the generic plot of most Hindi and Malayalee serials. Patricia (leader of a women's self-help organisation) referred to the Malayalee evening serials as sagas of 'weeping women'. Usha (2004:10), who interviewed women about television viewing in 200 households in Trivandrum and a nearby village, wrote that Asianet and Surya 'vie with one another in getting women glued to their programmes by producing and telecasting absorbing sob-stuff. Crying heroines and their tribulations dominate the serials'. The next section will explore what they are crying about.

Watching serials with people was no easy task. People did not want to seem overly interested in what they feared would be regarded as 'frivolous' entertainment. Also, I had to deal with the 'CNN syndrome': no matter how much I insisted that the family I was visiting watch what they usually watched, someone would eventually flip the channel back to CNN because they thought it would please me. One evening towards the end of my first year in Trivandrum, I decided to get around this problem by blatantly assuming the identity of researcher. I took a notepad to the home of Bala (Hindu Ezhava) and his family (described in Chapter 5), and with the assistance of Rija, Bala's 24-year old granddaughter, took copious notes on the programmes, advertisements and the ways Rija, her 20-year old sister and their grandmother watched and reacted to the serials.

## An evening at the serials

On this evening, I arrive at Rija's house at 1900. Rija takes a seat beside me and turns the television channel to Surya television, the Trivandrum-based channel that carries the serials she and her family watch every evening. Rija tells me that virtually all of the women in the neighbourhood (Kannamoola) have 'at least one eye' on the serials that run from 1900 to 2200 Monday through Friday; one eye, because many women, especially working women like Chavita (whose day was described in Chapter 3) are cleaning, cooking, putting children to bed or doing other household chores. Rija says (with a smile) that the men in the neighbourhood complain because they cannot get a cup of tea between 1900 and 2200, when the last serial has ended.

### *Sthree Malayalam* (Malayalee women)

The first serial is entitled *Sthree Malayalam* (Malayalee women). The central figure is a young woman named Meaman, one of the first 'weeping women' I will encounter on this evening. Rija explains that in previous episodes of *Sthree*, Meaman's mother died, her father remarried and then he died when Meaman was still a teenager. Her mean-hearted stepmother made Meaman's life miserable. The most dramatic conflicts had to do with Meaman's passion for singing and her interest in pursuing a singing career. The stepmother regarded this as foolish and immoral, and put obstacles in Meaman's way. When she turned 20, Meaman finally escaped the grips of her stepmother and moved in with her uncle (her deceased father's brother). He is a warm and understanding man who encourages her to pursue her interest in singing. In the meantime, Meaman falls in love with the neighbour's son Babu. Tragedy awaits because Babu has two unmarried sisters. He feels it is his duty to find husbands for them before he can even consider marriage. He tells Meaman it may take years to accumulate enough money and goods to support dowry for his two sisters. At this point, I ask the assembled women what caste the families in the serial belong to. Rija, her grandmother, and her sister give the impression that this has not occurred to them. Then they all agree that caste is not a central theme except in plots that focus on the dilemmas of cross caste marriage. Another popular subject in marriage conflict is social class. In *Sthree Malayalam*, Meaman's family is wealthy, while Babu, the neighbour boy, is from a lower middle class family. This class difference poses a barrier to Meaman's marriage prospects. Neither Meaman's stepmother nor her uncle is likely to approve her marriage.

At the first commercial break (after only seven minutes of programming), Rija comments that the usual pattern the serials is this: the central female character is confronted with one dilemma after another, some of her own making and others which are forced on her. The stories are constructed around the ways that women deal with these situations. According to Rija, a moral buried in most stories is that the heroines are those who stick to their duties as daughters, wives and mothers, no matter how severely they are tested. Rija observes that serials draw on the popularity of the story of Sita in *Ramayana*, who epitomises this moral.[2] The story of Sita was itself serialised on Indian television in the 1990s and remains one of the most popular programmes ever produced in India (Appadurai 1996). Rija believes that the character of Sita is the prototype for the heroines of many modern Malayalee serials.

This first commercial break lasted four minutes and contained advertisements for 13 products in the following sequence:

o Fairness oil
o Ayurvedic shampoo
o Biscuits
o House paint
o Panty hose
o A powdered additive to milk (to make a flavoured drink for children)
o Body soap
o Ceiling fan
o Chocolate bar
o House paint
o Toothpaste
o Diapers
o An *Ayurvedic* medicine for sore joints and muscles

Rija tells me that advertisements for fairness oils like the one in this sequence dominate prime time advertising (see Chapter 3 for a discussion of the importance of fairness). Fairness oils are very popular among Rija's friends and university schoolmates. She says that Kerala is known around

India for the quality and variety of its fairness products. Her family in North India recently asked her to buy and mail them four bottles of one of the locally produced *ayurvedic* fairness products. This is because one of Rija's cousins is starting to get marriage proposals. Rija tells me that fairness was itself the subject of a popular serial two years ago. The main characters were two sisters, one light-skinned and popular, the other dark-skinned and unpopular. Each episode was built around a dilemma of the dark-skinned sister and a triumph of the light-skinned one.

After the advertising break, *Sthree Malayalam* continues. The scene moves to another of Meaman's neighbours, a family consisting of an aging father and his middle-aged son, both heavy drinkers, and his two daughters, both in their early 20s. On this particular evening, the son has brought a woman home who he claims to be his new wife. Knowing this is a lie, his sisters heatedly insist that he and his girlfriend leave. The brother ignores them and retires to the bedroom with his 'wife'. In the meantime, the father is getting very drunk in the living room. The daughters enter the living room and find him passed out. They make a cursory effort to revive him, but thinking he will sleep it off as usual, give up trying and go out to visit friends. What they do not know is that he has fallen and knocked himself unconscious.

The second commercial break, after six more minutes of programming, lasts four minutes. Fourteen products are advertised:

o Fairness oil
o Female tampons
o Fanta
o Clothes washing powder
o Candy bar
o Gold watch
o House paint
o Clothes washing detergent
o Ghee
o Chocolate candy
o Toothpaste
o Bicycle
o Mango juice
o Body soap

The serial resumes with the daughters returning home late in the evening to find their father on the floor with blood in his hair. They try but fail to wake him; so, they call for help and manage to get him to the hospital.

After examining him, the doctor tells them that their father has had internal bleeding and his chances of living are only 50-50. The girls agonise. The final scene fades with the daughters crying despondently.

Rija comments that the moral to be drawn from this sequence is that no matter what the circumstances, nor how disreputably family men conduct themselves, the women of the family should remain dutiful to the men of the family.

Two products are advertised before the final credits:

o   Fairness Oil
o   Body soap

### *Sthree oru jwala* (Woman, Aflame)

The title of the serial beginning at 1930 is taken from Sita's trial by fire. The story centres on life in a rural village. In this evening's episode the main character, a teenage Hindu girl named Chaki, faints at school. The school nurse examines her and becomes suspicious that Chaki is pregnant. When she informs Chaki's teachers about this, they are shocked. Chaki has the reputation of being a sweet and dutiful girl and is one of the best students in the class. The teachers take Chaki to the village clinic, where the doctor confirms that she is pregnant. The scene switches to a close-up of a teenage boy. By his distressed look, we know he is Chaki's lover. We can also see that he is a Christian.

There is a commercial break after six minutes of programming. The 12 advertisements are almost identical to those in the first serial, as are the breaks and commercials which follow.

Back at the school, the teachers decide they must inform Chaki's father about her situation, but before they do anything they decide to seek the advice of a retired teacher who is a friend of the girl's father and highly respected in the community. Since Chaki's suspected lover is a Christian, the father's friend decides to consult with the village's Christian priest before going to Chaki's father. The two of them discuss the situation and then proceed to the home of an elderly Muslim man to seek his advice. Rija interjects that she thinks that the inclusion of the Muslim in the plot is an effort by the producers to be politically correct by including a wise man from each of the major religions.

The three wise men go to see the father, but are told that he has just recently been admitted to the hospital with a serious heart problem. The penultimate scene leaves the trio of wise men, and us, the viewers, contemplating whether or not they should risk giving the father the traumatic news about his daughter. In the brief final scene, the village gossip, who often stirs up trouble, watches Chaki leave the clinic with her two teachers. The programme ends leaving us certain that he will stir up trouble.

This ending stimulates a discussion between Rija and the other two women about the village gossip. Gossips figure often in the serials. The gossips are usually unsympathetic characters that often draw attention to the 'social rules' and how violators of these rules will almost always be held accountable. Otherwise, the story line takes advantage of a popular theme in serials and Bollywood movies, the play between passion and duty; between romantic love and affection; between love marriage and arranged marriage. At this point early in the story development, we do not know how the plot or the moral will develop, but according to Rija, it will likely conform to the usual scenario. The boy and girl, innocent and duty bound as they may be have made a mistake and succumbed to passion. Tragedy in one form or another likely awaits both. If their punishment is to be tempered, it will be through the wise intervention of the village elders.

### *Sthreejanmam* (Women's Lives)

The serial at 2000 is the longest playing Malayalee serial, entitled *Sthreejanmam* (Women's Lives). The main character is a woman named Lakshmi whose life is yet another that has consisted of confrontations with one dilemma after another. According to Rija, earlier episodes centred on the dramatic events surrounding Lakshmi's marriage. Her family had arranged her marriage to a man who Lakshmi knew and liked. However, just before the marriage ceremony, she was kidnapped (again, reminiscent of Sita's fate in the *Ramayana*). Her kidnapper held her captive for some time and then forced her to marry him. As Rija explained it, Lakshmi went along with the marriage because the kidnapping had already spoiled her reputation; hence, she could never attract another husband.

Soon after the marriage, Lakshmi has a child. The husband suspects the baby is not his (once again taking inspiration from the *Ramayana*). His suspicions deepen when he finds out Lakshmi has reestablished a friendship with her former fiancé. The husband becomes violent with both Lakshmi and their child. The neighbours report him to the police. He is arrested, tried and found to be mentally unstable. He is committed to a mental institution. In later episodes, the institution's doctors tell Lakshmi that her husband has no hope of improving unless he is allowed to see his son, who by this time is about five years old. After much anguished deliberation, Lakshmi takes the boy to visit his father.

When they arrive in front of his cell, the boy is afraid and resists approaching his father, but Lakshmi forces her son to embrace his father through the bars of his cell. When the boy refuses, the man collapses. In the last scene, Lakshmi is shown crying despondently while the orderlies try to revive him.

I ask for Rija's interpretation of the story's moral. She sees it as another variation on the same theme: a woman's duty is to stand by her husband no matter what the circumstances of the marriage or how she is treated afterwards. We are made to sympathise with the kidnapper-husband as much as with the dilemmas of Lakshmi and her son.

The three commercial breaks in *Streejanmam* each consist of advertisements for ten products and last about four minutes. All of them follow roughly the same sequence of the first commercial break, which is:

o  Kaveri, the most popular fairness cream based on ayurvedic principles.
o  Maruti car
o  A toy (the mascot of Cartoon Network)
o  Toothpaste
o  Facial soap
o  Candy
o  Gold watch
o  Body soap
o  Cookies
o  Candy bar

A few minutes after the end of this serial, around 2035, there is a power outage. By 2100 the power has still not returned. I call it a night, thank Rija and the other women and cycle home.

Rija and her family follow the serials because they find them to be engaging and entertaining. They are popular sources of gossip and discussion. Rija told me that her aunts, grandmother and their neighbours talked about the characters and their dilemmas in ways that resembled their gossip about real families. In this sense, the serials are 'active and socially negotiated', as Wilk found local programming to be in Belize (2002a:174). Usha (2004), who studied television viewing in Trivandrum, writes that her interest in the impact of serials grew out of this same blending of real and imaginary characters in her neighbourhood in New Delhi. She was puzzled when the name Hari kept coming up in discussions among neighbourhood women, because Usha had never heard of

Hari. When she finally asked who this person was, she was told that Hari was a character in a popular Hindi serial.

The themes of these serials draw on many of the same dilemmas for women explored in Chapters 3 and 4. The heroines are made to master multiple social pulls and unjust treatment, they inevitably conform to what Usha calls 'the epitome of the "male ideal" (Usha 2004:14)'. Concerning a typical heroine from the Asianet serial *Sthree*, Usha wrote, she could be portrayed as 'educated enchanting, beautiful...yet she would not fight for her own rights and privileges as a wife, simply because it would be inconvenient for the husband, to whom she believed, she was duty-bound to obey (2004:14)'. Usha found the Malayalee projection of the ideal woman to be much more conservative than those in serials in Tamil Nadu and Karnataka, where women characters display 'a certain inherent vitality and independence quite alien to the Kerala woman (2004:13)'.

These Malayalee programmes feature female characters who attempt to find a balance, described in Chapters 3 and 4, between being modern and yet traditional. Unlike a serial like the North American 'Dallas' or the Hindi serial *Stri*, which are centred around the lives of the rich and premier their material wealth (cars, villas, extravagant clothing), Malayalee serials do not premier the lives of rich families or the social advantages of conspicuous consumption. The messages in the serials do not associate the consumption of new commodities with better lives; however, while these popular programmes themselves do not encourage new ways of consuming, there is no doubt that they assemble a large and attentive audience for advertisers.

## Television advertising

The term 'soap opera' came from the potential of programmes so named to grab and hold an audience's attention and thereby allow advertisers to sell them soap during commercial breaks (Gledhill 1997). My 90 minutes of viewing serials with Rija reveals the tremendous volume of advertising, some of it involving soap and facial creams, but also many other kinds of products. On average, during evening programming, one product is advertised for every 30 seconds of air time.[3]

Television advertising is effective because it has the advantage of drawing on a wide range of communication forms and styles. 'It does not have the abstract or solitary quality of reading and writing ...but rather shares something of the nature and impact of the direct personal interaction which obtains in oral cultures (Goody and Watt 1968:69).'

Television advertisements are much more than just information; they have a purpose, which is to change the way the viewer consumes. Falk claims that this use of advertising to actively sell and market is a characteristic of the advent of 'modern advertising' after the mid-19th century. Prior to that time, advertising was simply information and advertisements were used to inform potential customers about the 'existence and availability of a certain product'. From the 1950s in North America and Europe, the purpose changed. From that point forward, the purpose has been to 'stimulate demand and thereby to sell as much as possible' (Falk 1997b:65).

Exposure to Indian television advertising leaves no doubt as to its purpose: to convince viewers to buy products. The plots, themes and use of characters are sophisticated and play on insightful social themes. Social inclusion and distinction are two common themes in Kerala television advertising. Advertisements demonstrate how the advertised product can help the consumer to either stand out or fit in, or do both simultaneously. As Jhally writes, 'advertising thus does not work by creating values and attitudes out of nothing but by drawing upon and rechannelling concerns that the target audience (and the culture) already shares (2002: 329).' At the most superficial level, advertisements attempt to get viewers to replace the brand or model of a thing they are using (soap, washing machine, mixie) with a different one. A more ambitious goal is to introduce a consumer to an entirely new product category such as a microwave oven, mobile phone or car. An advertisement is designed to show how this *new* product will make life better, easier, more comfortable or hygienic. Falk (1997a) sums up the purpose of advertisers this way: 'There are various different ways to argue in favour of a product; you may say it is "useful", "comfortable", "healthy", that it brings "social prestige" or simply that it "makes you feel good". The crucial thing is that an image is created of a "good object" – that *you* do not yet have.' Appadurai (1986:56) writes that a longer term objective of advertisers is to not only sell products, but also to transform consumers: 'The images of sociality (belonging, sex appeal, power, distinction, health, togetherness, camaraderie) that underlie much advertising focus on the transformation of the consumer to the point where the particular commodity sold is almost an afterthought.'

In the 1990s, TNCs and international advertising agencies brought their long experience with transforming consumers to India and Kerala. Nonetheless, many TNCs experienced much lower sales than anticipated in the early and mid-1990s. Both the TNCs themselves and independent analysts attributed this to flawed marketing strategies. According to

Mazzarella (2003), the key to the marketing successes of those TNCs who were most successful was the creation of marketing images that took advantage of the products foreign origins (its distance), yet made it a natural part of Indian practice (closeness). Successful advertising images are those which remake the global as local and modern as traditional. Mazzarella outlines the competition between Coca-Cola and Pepsi for market share in India in the 1990s. Pepsi was successful in its marketing because it did not rest on the laurels of its brand recognition, but rather gave its image an Indian flavour. It created 'maximum spectacle' with music, dancing and the use of Indian celebrities (2003:220). In a typical Pepsi advertising plot, the actress drinks a bottle of Pepsi while at the same time she throws caution to the wind, dancing, moving, dressing and acting in ways that draw the attention and awe (at her boldness) of classmates, friends and passers-by. Coca-Cola, which also aims its advertisements at youth, initially simply directed its global advertising message to Indian audiences. The message was that of solidarity and caring, portrayed by showing multi-ethnic groups of teenagers holding hands and singing. This message flopped with Indian teenagers (Liechty 2003). Coca-Cola went back to the drawing board and adjusted its strategy. It began using Indian actors and sports celebrities. Aishwarya Rai (the Miss World winner discussed in Chapter 3) was the star of many of these. Her poses, dress and behaviour involved a mix of Western and Indian looks and styles. Coca-Cola advertising succeeded with this variation on the theme of 'close-distance' embodied in the look and behaviour of Aishwarya Rai and other celebrities. Coca-Cola's sales began to catch up with Pepsi's in the latter part of the decade of the 1990s.

The use of the close-distance concept is also evident in the advertising of household appliances. In many appliance advertisements the setting or context of the advertisement reflects the Indian 'traditional' while the household appliance represents the distant 'modern'. The idea is that the owner does not disturb their 'Indianness' by possessing the appliance; on the contrary, ownership is portrayed as a characteristic of the modern Indian. This is similar to much of Japanese appliance advertising in the 1950s and 1960s, the period in which virtually all of the North American household appliances made their way into Japanese homes (Wilhite et al. 1997). An example is a series of advertisements for Mitsubishi air conditioners. The image consisted of a traditional Japanese room, with tatami floors, traditional decoration and a woman in kimono kneeling in the foreground. Prominently visible on the wall behind her was a Mitsubishi air conditioner. The intention of the image was to recast the normal Japanese home as one which included an air conditioner.

In its India advertising, Carrier (a US-based air conditioning manufacturer) uses this technique of upgrading the notion of traditional to include a new commodity. In a Carrier air conditioning advertisement from the 1990s, the advertisement opens with a scene in which a holy man (*sadhu*) is in the foreground (Fernandes 2000). He is lying on a bed of nails in a hot, dirty and dusty street. His prone body remains in focus as the background is abruptly changed. He finds himself laying comfortably on a bed in an air-conditioned bedroom. Fernandes writes:

> In both photographs, the *sadhu* is depicted in exactly the same position, reclining with his eyes shut. The image provides a visual embodiment of the ways in which the emergence of Indian national identity rested on the preservation of a protected inner sphere of tradition...The core of Indian tradition, the image suggests, can be retained even as the material context of that tradition is modernized and improved.

This same type of transformation is also evident in advertisements depicting the Indian housewife. Usher (2004) did an extensive analysis of the way women were portrayed in television advertisements and found that housewives were portrayed as 'smart in apparel and appearance, and shrewd and efficient with regard to disposition of time and management of finances (2004:22)'. This is supported in Mankekar's analysis of Indian serials of the 1990s and their advertisements (1999:91):

> In these serials and these advertisements, women were depicted as simultaneously modern and traditional: even as they ran their homes with modern appliances, they were portrayed as traditional because of the fidelity to their roles as dutiful housewives and nurturing mothers. In the process, the advertisements *constructed* (original emphasis) the meanings of 'modernity' and 'tradition' in explicitly gendered terms.

Concerning the role of men, Usher writes that in Kerala, 'In the world of advertisement, women provide the humble services while man provides the useful advice (2004:20).' The message is that modernity in the form of commodity ownership can be achieved without threatening either the role of husband or wife.

Advertising may be voluminous and sophisticated, but how effective is it in changing consumption? Based on my study, there is no doubt that people are interested in television advertisements. Very seldom does anyone use the commercial break as an opportunity to leave the room. People are interested in their plots and curious about new products. An elderly Hindu Brahmin lady, who saw herself as religious and was rather austere in her lifestyle, surprised me when she told me, 'I watch TV because of that (advertising). I love watching them (advertisements). And sometimes it helps me, if I want to buy something.' Abraham (Syrian Christian) told me that he was always on the lookout for 'new' products. He scanned advertisements and was occasionally inspired to buy something. As an example, he said he purchased a Honda motorcycle after an oft-repeated Honda advertisement that had caught his eye. His wife Mea, who was present when he told me this, added, 'When they (advertisements) are catchy we stop to see what the whole thing is about.' As these quotes suggest, I found little evidence in Kerala of the cynicism that characterises many peoples attitude to advertisements in Europe and North America.

Another more direct piece of evidence that advertising affects consumption is that many people attributed their purchase of a specific product to an advertisement they had seen. Abraham's purchase of his Honda is one example. Sindu (Hindu Nair) told me that after watching a washing machine advertisement she went out and bought precisely the machine advertised. Many parents told me they had purchased something that their children had seen advertised. An example is Shobana's (Hindu Ezhava) purchase of their first refrigerator. Her children had seen a television advertisement that emphasised its capacity to keep bottled drinks cold. Shobana said they were interested in having the potential to offer visiting neighbourhood friends a cool soft drink. Shobana's sister told me that the boys were constantly pestering Shobana to buy things they had seen advertised. Many of these were 'Cartoon Network' advertisements, which heavily advertised ready-made food, toys and clothes. Lathika, a Hindu Nair mother of two children, told me that she had cancelled their cable connection because of her children's zealous interest in 'Cartoon Network' and its advertising (Sateesh was another who made the same choice, the social consequences of which are discussed in Chapter 7):

> She (her 10 year old daughter Lakshmi) wants most of the things which is shown in the advertisements. If there is an advertisement of a shoe, then they (Lakshmi and her friends) want that

shoe, if it is a school bag then they want that. Dresses and sweets and also bakery items which are shown in the advertisements, they ask for.

Children's advertising is not only effective in Kerala. Varma (1998) cites an all-India study of children's television viewing from 1996, which found that 75 per cent of children wanted to possess the products they had seen advertised on television.

Store managers and salesmen of automobile dealerships and household appliance stores were convinced of the power of television advertising. Several said that their clients had come into their store asking for a specific car or appliance that they had seen advertised, including its colour and other details. In the 408 structured interviews, 21 per cent of participants indicated that they had purchased something they had seen advertised on television. Those products most often named were soaps, cleaning products, cosmetics and toothpastes. Usha found that in over 50 per cent of the 100 Trivandrum households in her study, participants 'openly admitted to being influence by the commercial advertisements in their choice of consumer products like detergents and household goods (2004:22)'.

My assessment is that Malayalee are interested in advertisements for their informational content. Some people are inspired by advertisements to buy certain products. However, they are far from 'duped' by advertising as some analysts contend (see Nava 1997). A young Hindu Ezhava father told me, 'We never feel we *have* to buy some thing by watching the TV advertisement. But sometimes when we go to buy something from the shops and see some other thing which has been on television, we feel to buy those.' Savita, a single 22-year old Hindu Ezhava told me, 'I guess you wouldn't fall for a product just because it is continuously advertised, but when you go to the market you always identify the product and you might test the product and if you like it you give it a go.'

## Conclusion

Television's 100 channels bring a diverse set of ideas into many middle class Trivandrum homes, some of local and others of diverse foreign origin. To understand the importance of television programming in Trivandrum after the mid-1990s, it is important to keep in mind that the most popular channels are local and the most popular programmes are produced locally. The messages, themes and metaphors which these programmes convey grow out of the local context and most often treat subjects that are of local concern. These include such things as gender roles within family, generational conflict, marriage practices, as well as social rules for conduct and the consequences for deviating from them. Their themes do not regularly or systematically premier consumption as a path to a better life. However, foreign-based channels and international programming popular with young people have provided glimpses of the ways some of these same topics are played out elsewhere in the world, or as Abu-Lughod (1999:123) puts it, they are a source for 'intensifying and multiplying encounters among life worlds, sensibilities, and ideas.' These ideas from viewing, and from reflecting about what they have seen are bound to lead to *some* changes in the ways *some* people think about consumption; however, the changes are subtle and the relationship between globalised media and local practice is difficult to pin down.

The effects of advertising are easier to identify. There has been a massive increase in the amount of advertising in Kerala over a short span of time. The audience is interested and receptive. The amounts spent on advertising are enormous and the themes clever, playing on skin colour, exclusion, normality, efficiency and others which go to the heart of conflicts and negotiations concerning Malayalee social identity. It would be impossible to determine conclusively the extent to which advertisements have been successful at doing what they are intended to do. However, based on the interest in advertising, the attention paid to it and the fact that many people could point to something they had bought because they or one of their children saw it advertised, there is ample support for the claim that advertising is an important contributor to changing consumption in Kerala.

# 10
# Conclusion

It should be clear from a reading of the preceding chapters that there is no single theory that captures the ways that consumption is changing in South India over the past few decades. Changes are implicated in the precipitous change in India's relation to the global economy, in family and kin relations, as well as in the embedded potentials of new products and technologies. Two different forms for globalisation are relevant to change, one being the opening of India to transnational capital and its agents, the other being the transnationalisation of work and residence. However, equally important are the socio-cultural contexts of consumption practices in South India, including the relations within the extended family and the endogenous history of the house and home. Understanding these changing contexts is essential to understanding how and why people acquire new goods and how they become normalised in consumption practices. This study has revealed that the individualised, inner-directed conceptualisation of the consumer is not relevant for understanding many forms for consumption in South India. Consumption invokes family, and family networks are important to acquisition of goods. Since the south Indian extended family is often spread over great geographical distances, the consumption of goods may begin abroad, pass through several phases in their cultural biographies, and end up in homes in Kerala (Werbner 1990 and Kopytoff 1986). These complex forms of consumption demand a rethinking of consumption theory for Southern consumption, and offer new insights which are relevant for theorising consumption in the North.

## Transnational agents

One of this book's most important points is that changing consumption in India is not just about 'opening' to the transnational agents that were barred from India for much of its early history. Nonetheless, their contribution to change should not be underestimated. Multinational enterprises and their products are the powerful bearers of new ideas about consumption. The multinational actors in the form of TNCs and international advertising agencies draw on considerable resources and experience. As Mazzarella (2003:18) writes in *Shoveling Smoke*:

> Transnational marketing and advertising agencies, commercial mass media, and all the auxiliary services that accompany them – are perhaps the most efficient and successful contemporary practitioners of a skill that no one can afford to ignore: namely, the ability to move fluently between the local and the global, as well as between the concrete and the abstract.

Indians since the mid-1990s are being inundated with marketing and advertising messages which positively value new products and new ways of consuming. This or that product will help people to fit in, or to stand out, or to do both simultaneously. The exposure to advertising in Trivandrum has changed on a massive scale over a relatively short time period. Though its effectiveness is difficult to ascertain, it is fairly evident that advertising leads people to reflect about their consumption of things like cleanliness, beauty and comfort. There is no doubt that people pay attention to advertisements and regard them with interest and curiosity. Advertising is viewed by many people as an important source of information on new products and how they can enhance their lives. Whether people will maintain their interest in advertising over time or whether advertising will become so commonplace that its agentive power declines remains to be seen. In any case, this study joins those of Liechty (2003), Mankekar (1999), Burke (2003) and Nava (1997) in making the claim that advertising and other forms of commercial discourse are important to theorising consumption change in the South.

The everyday life of south Indians is not only exposed to massive increases in commercial advertising, but to a change in urban landscapes which are increasingly filled with billboards, show rooms, retail outlets and shopping centres. This is similar to developments in Nepal, which Liechty writes has been subjected 'to an enormous increase in the quantity of commoditised forms and their sudden ubiquity in daily life'. In India, entirely new kinds of products inhabit the spaces of everyday life, with their origins not only abroad, but also as a result of changing production in India. Concerning thermal cooling, we have seen how changes in building materials and house designs have steered consumption choices in the direction of big houses that perform like thermal collectors; as a result, air conditioners are needed in order to make them liveable. In the domain of food consumption, transnational investment has influenced local production of both refrigerators and washing machines. One result of this transnational influence is that the size of refrigerators produced in India has grown significantly since 1990. Production of the semi-automatic, compact washing machine – a best seller in the 1980s – has decreased dramatically. This is one demonstration of how production structures consumption choices. In India the set of things that consumers have to choose from has changed significantly over a course of 15 years. As consumers seek new products they may find that familiar products are no longer available, or that products have materialised to satisfy wants they never knew they had. Demand is not always a magic wand that conjures up goods, or in the words of Brewer and Trentman, 'Consumer goods and services are not the product of immaculate conception (2003:3).'

The combined forces of transnational capital and transnational development are important agents in changing production in India. Organisations such as the World Bank and the Asian Development Bank have embedded new products and new ideas about consumption in their development programmes. These multilateral development actors are powerful change agents because they couch new forms for consumption in the language of development. Cleanliness is an example of a global discourse that has brought the full power of the global development capitalism to bear in India. Discourses about beauty, health and other forms for cleanliness are made more powerful because of their association with modernity, progress and development. In Trivandrum, we have seen how woman – and her body, appearance, home and family – are the targets of many of these discourses.

In recent years, multilateral development agencies have become increasingly involved in promoting changes in consumption from the perspective of environmental amelioration. My impression is that these programmes suffer from the same lack of attention to local practices as does the soap campaign discussed in Chapter 7. Implicit to many of them is the assumption of a globally generic home and generic consumption practices. Those programmes which have focused on energy efficiency and the reduction of climate-change emissions, such as the multi-lateral and bilateral technology transfer programmes supported by the United Nations' Clean Development Mechanism and Joint Implementation initiatives, suffer from these assumptions about the generic home. At the heart of many of these programmes is a form for technology progressivism. One aspect of this is a presumption that new, modern technologies can be made to 'leap' from Europe or North America into everyday life anywhere, increasing the efficiency of achieving a given energy service (such as comfort, cleanliness, food or mobility) without having ramifications for other practices.[1] Very few of these global initiatives have made any serious effort to incorporate local knowledge or to take account of, and build on local practices. The end result is that while the technical energy efficiency of household appliances has been improved, the total energy use associated with home cooling, food preparation, and clothes washing has actually increased (Wilhite 2007). As the examples of refrigeration and air conditioning (in Chapters 3 and 7 respectively) indicate, the 'leapfrogging' of efficient technology devices into local consumption practices has, and could continue to result in an increase in the total consumption of energy if there is not a better accounting of the local socio-cultural contexts into which the technology leaps.

## Social relations and cultural practices

This book has recounted the ways in which cultural practices such as those related to appearance, religion, marriage and dowry are important to consumption and to the ways that consumption is changing. The vast numbers of new beauty and cleanliness products available in India, and their marketing draw on traditional ideas about fairness and purity. Fairness has been commodified in a sophisticated array of new packaging and products. At the same time, colour cosmetics owe much of their popularity to their association with freedom and liberation from gender binds to housework and home. On the other hand, the traditional practice of dowry has thrived on new forms for consumption, now encompassing significant amounts of cash, gold, and, increasingly, cars and household appliances. Dowry in south India can be seen as a form for consumption that involves new commodities, marriage practices and the exercising of extended family networks. An important insight from this study is that Kerala dowry is consumption and that both dowry and consumption have been strengthened by the engagement of global markets with traditional practices.

Another related insight from the study involves the importance of family in consumption. The results of this study constitute a contradiction to much of the theorising about consumption in the West, in which family and family relations have virtually dropped out. A consequence of this is that family has not been given much attention in recent studies of consumption in the developing world. Studies such as those of Liechty (2003) in Nepal, O'Dougherty (2002) in Brazil and Colloredo-Mansfeld (1999) in Bolivia, put middle class identity at the forefront. For example, Liechty emphasises that consumption in Kathmandu is tied to the project of constructing 'class cultural lives (2003: 255)'. O'Dougherty writes, 'Appropriation of certain things is clearly central for claims to a class identity at or above the level of the individual or family in question (2002).' I claim that the reverse is true in Trivandrum: family is at a level above class identity as a reference for consumption. Gender relations, work, marriage and migration all implicate family in consumption. Family deserves renewed theoretical attention in consumption studies in the South. This is a view supported by Daniel Miller (2001a) for studies in the North as well.

To turn to another important point concerning family and kinship, changes in the ways the joint family is organised into households has been important to changing consumption. I initially interpreted the rise of the nuclear household as evidence for a weakening of the extended family. However, I was to discover that in spite of being segregated into nuclear households, the extended family still constitutes an important social force and social network. Virtually every newly married couple in Kerala begins their married life in the joint family household and continues to be a part of it until the first child is born. During this period, family traditions for child care, cooking and other important practices are passed from one generation to another. Once established in their separate nuclear families, physical contact with the new household's parental generation diminishes, but family bonds remain strong. Consumption is an important factor in reinforcing ties across these separate households through family dowry, wedding gifts, and the flow of goods and money in Kerala workscapes.

The transition from joint to nuclear households over the course of a few generations has what could best be called a structural consequence for consumption. As joint families are segmented into nuclear households, the need for furnishings multiplies and the total space used to house the family increases. Whereas previously 8–12 family members shared a single living space and one set of household appliances, today in a Trivandrum nuclear family 4–5 people, on average, share the home and its furnishings. Thus the half-century-long break-up of the joint family has led to an increase in consumption of building materials, appliances, furnishings, and per capita housing space. This increase has in turn resulted in an increased demand for mechanical cooling, water and energy. Elsewhere in the South (India, parts of Asia and Africa), the joint family household is still the dominant living arrangement. The segmenting into nuclear households will likely continue in these parts of the world, contributing to an increasing consumption of house and appliances. In the rich countries of Europe and North America a new phase of household segmentation is underway. Aging populations and frequent divorces have lead to segmentation of households into smaller units (reductions in average household size), which leads to further structural increases in consumption.[2]

## Technology

Thus far the focus has been on the way things are acquired, but consumption also involves how things are used. As mentioned above (and developed in Chapters 4 and 7), the use of technologies such as washing machines, air conditioners, and televisions has wide ranging and sometimes unanticipated effects on consumption practices. Whether they are produced locally or distantly, and whether they be simple and transparent or complex, technologies have embedded knowledge and this in turn influences practice (Bourdieu 1998). Thus, technology ought to be given greater attention in studies of everyday consumption. Unfortunately, technology agency has not drawn much interest in anthropology (Sigaut 1994), even though one of the most influential anthropologists of the early 20th century, Marcel Mauss argued that technology ought to be a central anthropological subject (Pfaffenburger 1988). He made the point that technology is a product of human choices and social processes. These get embedded in the form of practical knowledge and codes for its usage by others. These embedded potentials are important in theorising why consumption practices change; however, while goods, material culture, commodities, and consumption, have all gotten considerable attention in anthropology, technology has not been in focus in ethnographic studies of consumption.

Why this aversion to technology? One reason is that anthropology's principle subject, the 'other' represented by the non-western, rural and non-elite peoples of the world, has not possessed advanced, complex technologies – such as the household appliance technologies – in significant numbers, at least not until recently. Another point is that many anthropologists would really prefer that technologies go away, both literally and epistemologically. There is much truth to Nigel Barley's (1986) semi-satirical depiction in *The Innocent Anthropologist,* in which the subject villagers, warned of the anthropologists approach, cast off their jeans, don their traditional clothing and pass their television sets out the back windows of their huts. This image has a double meaning of course, one being that the people we study are quite reflexive about what they think anthropologists want to see. However, it also suggests that many anthropologists would prefer to view technologies and other products of Western materialism as somehow exogenous to local worlds and even as a threat to them. The more glamorous media and information technologies *have* attracted fledgling interest in anthropology, examples being Mankekar (1993), Abu-Lughod (1999) and Wilk (2002a). However, the more mundane household appliances have received little attention.

This study joins ethnographies from Asia, Africa and other parts of the so-called developing world in suggesting that people want the everyday technologies of the West and are acquiring them in increasing numbers. In Kerala, they want them for the promised conveniences, comfort and for the entertainment they provide; and because access to them is seen as a form for righting global injustices: Why should the imbalances of an earlier colonial era be repeated, namely that the peoples of the rich countries have access to goods and while they are denied? Access to goods, including the everyday household technologies of the West is seen as evidence of the release of Southern consumers from the repressions of colonialism and underdevelopment and of achievement of a new status as first class citizens of the global community. This is captured perfectly in the statement of one of our neighbours, who said that 'Before independence English people did not give us anything. They were only taking from us. They took from here and they produced the output there and sold to us. To get knowledge about the world and to go along with the movements of the world ...was a part of the freedom movement.'

The increasing interpenetration of global capitalism and its products on the one hand, and the interest in foreign-produced technologies on the other, make any ethnography almost anywhere in the world incomplete without an accounting of them.

Another reason that technologies deserve special attention is that they distinguish themselves from other goods in important ways. Appliances such as refrigerators, mixmasters, and microwave ovens are multi-functioned and their workings opaque to their users. As Rip and Kemp (1998) point out, everyday household technologies do not end at the electric outlet or exit pipe. Appliances are connected to regimes of technologies beginning in the walls of the house and extending to neighbourhood, city, national, and interregional infrastructures that deliver essential inputs such as energy and water, and that take away outputs like grey water. If any of these technologies should malfunction, an expert is needed to get them up and running again. In places like India, the repair and maintenance of possessions, activities that were once a source of pride, become a source of alienation.

Other branches of the social sciences have given more attention to technologies and modern infrastructures. In the early 20th century, phenomenologists such as Edmund Husserl emphasised the importance of knowledge generated by or attached to things in use, in contrast to the importance of cognition and perception. Their point was that technology should not be regarded as an inert object but as something embedded with agency. As he put it, things have their own kind of knowledge. In the 1980s, Dutch scholar Wiebe Bijker, along with British John Law and French sociologist Bruno Latour were the founders of a research domain called the Social Shaping of Technology (SST), which grew partly out of a reaction to technological determinism. This SST tradition has generated a considerable body of literature on how people exert agency on technology, or how technologies are domesticated by their users. Latour and others at Ecoles des Mines in Paris later developed a new strand of thinking, with it roots in Heidegger, which theorises agency as shared between technologies and humans. Looking at their work, in my view, Latour stretches the anthromorphising of technology too far, but nonetheless the idea that technologies, once in use in homes, can then affect consumption practices in unanticipated ways is important in interpreting changes in consumption practices. In recent years sociologists such as Warde (1997) and Shove (2003) have made 'inconspicuous technologies' subjects of consumption research in Europe and North America. This important topic needs greater attention in studies of consumption in the South.

## Consumption as performance

Many argue that 'modern' and 'modernity' have so many connotations as to render them useless as analytical categories (see Latour 1993 and Kolshus 2005). However, the uses of modernity and being modern are so pervasive in Kerala that they are unavoidable subjects in studies of social change. In public policy in India and Kerala, as well as in the activist agendas of non-governmental organisations, modernity is used as a cognate of progress and development. Becoming modern has been used to promote legal, social and religious reforms; for example, in agendas for changing caste, family and gender relations. It has also been associated with the 'opening' of India to global markets, and with a drift in Kerala away from a planned economy and towards a market-driven economy and a privatisation of public services.

At the level of household, being modern is bound up with social performance. A car is the most obvious product which people use to make statements about themselves. The signalling of modernity is one explanation for why cars and some household appliances are popular in dowry. However, in Kerala consumption the aura of modernity is short-lived. As goods become habitual parts of consumption practices, the performative aspects of consumption change subtly. This 'agentive power' of the drive to fit in (Comaroff and Comaroff 1992:28) becomes as important to changing consumption as is the pursuit of modernity. People consume as others do in order to avoid the label of being stingy, poor or out of touch. This represents a significant change in the social dynamics around materialism and simplicity. In Kerala of only a generation or two ago, fitting in involved projecting an image of thriftiness or frugality. One of the significant changes in middle class comportment over the past few decades in Kerala is the diminishing social importance of frugality. Thrift and modesty in consumption are no longer identified as characteristics of a proper middle class family. Thus a powerful source of social friction to the increasing tempo of new consumption is waning in Kerala.

Work migration contributes to the increased tempo of the normalising of new consumption practices. For work migrants, many of the modern appliances are mundane and taken-for-granted in their countries of work. The habituation to them contributes to the normalising of luxury consumption and to its transfer back to migrant's families in Kerala. Given that work migration and other forms for transnational residence are increasing around the world, this modern-normal change dialectic is important in theorising consumption change. In the words of Hindu Ezhava Govinda, 'The telephone, refrigerators, these were luxuries for me when I was younger. Now they have become common things and also necessary things... This is a fast changing time. There will be an end to this, but I don't know when.'

## Consumption and the environment

Consumption is viewed by most South Indians as a path to a better, freer life; it will bring social status, better comfort, increased convenience and new forms for mobility and entertainment. However, an unacknowledged consequence of these freedoms is a heavy environmental price, both in the form of strains on limited resources and in the form of pollution. Will the rapid environmental deterioration in India motivate changes in consumption? I see this as unlikely at least in the near future, either in the ways consumption is treated in national policy or on the part of consumers. In this section, I explore why.

First, at the national level, macro-political indicators of Indian economic growth (such as Gross Domestic Product) count increases in consumption as positive. At the same time, environmental deterioration is undervalued in macro-economic models (as is the case in countries in North America, Europe and East Asia). Second, at the level of household, there is little awareness of the links between consumption and environmental deterioration. Finally, for the vast majority of those who *are* aware of the environmental consequences of their consumption of things like cars, soaps and household appliances, environmental considerations make little difference when weighed against the perceived benefits of consumption. And, why should we expect Indian consumers to weigh the environment in consumption decisions when consumers in the rich countries of the world have not done so to any significant extent? Take automobility, for example; the automobile fleet in both Europe and North America is less energy efficient today than it was in 1980. Within the home, the efficiencies of many household appliances have improved, but increases in the numbers and sizes of appliances have overshadowed energy efficiency gains, so that the bottom line has been a gradual increase in energy used. The size of houses has increased dramatically in virtually every Western country over the past few decades, increasing the heating and cooling demands. Once again, while some gains have been made in device efficiency, the overall consumption of energy for home comfort, hygiene and mobility have increased in most OECD countries.

Concerning global commitments to combating climate change, the USA stands out as a barrier to serious efforts to come to grips with the environmental consequences of consumption. At the first global conference in which a climate change agreement was on the agenda, then President George Bush (the first) forcefully asserted the American position with words to the effect that the American way of life is not negotiable. His son George W. Bush has adopted a similar position and has refused to sign international climate conventions. Within the past few years, there have been efforts in some US states, notably California, to encourage renewable energy sources and to reduce the environmental impacts of 'conventional' energy sources such as coal-based thermal power. These efforts are focused on changing the nature of production (the ways energy is produced), not on changing the nature of consumption (the ways people consume). Indians are well aware of these halting efforts by the United States and other Western countries.

Another reason for a lack of attention to the environmental consequences of consumption is that Indian middle classes share with their Western counterparts the ability to buy their way out of the worst consequences of pollution. Take energy, water and waste as examples. Catering to urban elites and middle classes, India's strategy for years has been to place its highly polluting coal-based thermal power stations in poor, rural areas. Many local communities do not even have feed-ins to local electricity delivery systems. The power is sent through high voltage transmission lines to the big cities. Thus local people get the environmental consequences of having energy production facilities in their communities, but not the benefits. For the urban elites, the long transmission distance adds to the price of the power but deposits the pollution far beyond their purview.

Looking at another essential resource, clean water, the increasing middle class consumption of washing machines and soaps is contributing to decline in water quality. For poor people who are not connected into public water delivery facilities, they must search further and further a field for clean water. Middle class and elite families can use their economic advantage to purchase water from commercial vendors or to buy bottled water from retailers. Grey water runoff from middle class and elite neighbourhoods in urban settings like Trivandrum is much more likely to be carried away by a system of pipes and treatment facilities than in poor neighbourhoods, where infrastructures for waste runoff and treatment are non-existent. This contributes to pollution of water sources for drinking and cleaning.

Given the interest in new forms for consumption and the lack of interest or awareness in the environmental consequences, in my view, for India to make a dent in the growing environmental problems associated with consumption, it will have to focus 'upstream', using technology standards, regulations and commercial incentives to move the choices in any given product range in an environmentally-friendly direction. This will mean a broadening of the policy focus from technical efficiency to an engagement with the webs of technologies and consumption practices contributing to home services such as cooling, food preservation and mobility. The goal should be to find ways to use combinations of new technologies and local knowledge to provide these home services with a minimum environmental impact. Concerning the environmentally problematic developments in home cooling, the stage has already been set for deep changes in the ways people cool their homes. Nonetheless, there is still a potential in India to avoid locking into a path towards the air conditioned societies of North America and Japan. In order to avoid this development, policies will have to be put into place that encourage and reinforce the use of passive cooling where possible; in other words, to aim at retaining local solutions where possible and to support the development of innovations on house designs that facilitate non-mechanical cooling. In future consumption policies, the aim of favouring (or prejudicing) neither the local nor the distant is essential to encouraging change which is environmentally benign and socially just.

# Notes

## Chapter 1   Introduction

1. According to the India Census (1991), Kerala spent 50 per cent more on education than the average Indian state. Kerala's spending on health care in the 1990s was about twice the national average. As a result, Kerala has the most highly educated population in India and some of the most impressive health statistics. Kerala is first among Indian states in the provision of hospital beds and in doctor per population ratio. Its infant mortality rate is the lowest in India at 17 deaths per 1000 births compared to 91 deaths per 1000 births nationally (South India Human Development Report 2001).
2. The official name for Kerala's capital is Thiruvananthapuram. It is also known as Trivandrum.
3. This problem is not limited to developing countries alone. In Norway, for example, cited by UNESCO as the country with the highest living standard in the world, about a quarter of household waste is untreated.
4. Malayalee is the word used both by people from Kerala and elsewhere in India to refer to citizens of Kerala.

## Chapter 2   Global Interchange and Modernising Reforms

1. Prakesh (2003) outlines the importance of religious reform over the whole of India in the late 19th century, claiming that this was motivated by an effort to legitimise Hinduism and its proponents by stripping it of its 'superstitions' and 'myths'. Indian reformists attempted to recast Hinduism 'in the image of Western reason (2003:41).'
2. www.prd.kerala.gov.in
3. Gujarat is a state in Northern India.
4. The Mappila were low-ranking tenants, labourers and peasants. They battled against oppression by their Indian landlords and the British Colonial Authority many times between 1836 and 1921 (Panikkar 1989).
5. The Communist Party split into two parts, CPI and CPI (Marxist), soon after Kerala was formed.
6. The Nair, some members of the Ezhava castes, as well as other smaller castes such as the Samantans, Vilakkitallas (barbers), Veluthedatus (washer men) and, according to Fuller (1976), even some Muslims practiced matrilineality and matrilocality well into the 20th century. The Hindu Nambuthiri Brahmin caste, less than 2 per cent of the population, practiced patriliny, but the Nair matrilineal practice of *sambandham* allowed Brahmin men to have concubinal relationships with matrilineal Nair women. Christians and the vast majority of Muslims, who together constitute about 30 per cent of the population of South Kerala, have always practiced patrilineality and patrilocality.
7. Arunima (1996) contends that prior to colonial times it was not uncommon for a senior woman to assume the role of family *karnavar*. Colonial laws

recognised males as *taravad* managers and thus diminished the power of women.
8. According to Jeffrey (1992), Travancore courts required unanimous consent by members of the Taravad before the family could divide, sell or mortgage the property. A single dissenter was enough to invalidate a transaction.
9. The Vedic doctrine of Hinduism bases its principles on the *Vedas* (book of knowledge), the *Upanishads* (which deals with metaphysics), and the *Smitri*, the sagas and myths which deepen and explain the Vedas. These give guidelines for moral practice and family relations, as well as set out practices associated with birth, puberty, death and afterlife.
10. The Nambuthiri Brahmins is the name used by the Kerala Brahmins. According to Iyer (1981), it is derived from the words *nambuka*, meaning 'sacred', and *thiri*, meaning 'light'.
11. These were calculated after having eliminated from the sample households with reported incomes over 20,000 rupees per month. Approximately the same proportion of households were eliminated in each of the two castes: 22 of 146 (15 per cent) of the Ezhava and 14 of 98 (14 per cent) of the Nair.
12. In restaurants, the person sitting next to you is more likely to be an Ezhava than a Nair: 18 per cent of the Nair survey respondents indicated that they eat regularly in restaurants, whereas 50 per cent of the Ezhava indicated that they regularly ate in restaurants. As I will discuss in Chapter 4, one reason for this difference has to do with ideas about pollution. Members of the formerly upper caste Nair were concerned about the caste of food handlers in restaurants, while Ezhava were not.

## Chapter 3   Women in a Bind: The Crucible of Marriage and Dowry

1. According to Dube (1997), a similar change in dowry took place in neighbouring Tamil Nadu during the same period.
2. Sixty per cent of university undergraduates and 40 per cent of PhD students at Kerala universities are women.
3. These two appearance ideals are similar to those Johnson (1998) found for women in the Philippines, called *istyle* (traditional) and *adapt* (modern). Liechty (2003:74) found a similar duality in Kathmandu, where young women are expected to use make-up in subtle ways to enhance beauty; but as one of the women he interviewed said, 'Being *too simple* isn't good. But being really vulgar isn't good either. So one should be somewhere in the middle, it seems to me.'
4. The commodification of fairness is the subject of a historical novel by Davidar (2001). In it, a young South Indian doctor's experiments lead to a fairness cream with miraculous properties. The novel was very popular among our middle class acquaintances in Trivandrum.
5. In a typical Indian journalistic style, the author uses 'plumping for shape' to encapsulate the irony that young women are starving themselves to get thin, but at the same time eating candy and snacks because their starvation diet makes them hungry.
6. The *Times of India* listed the following contests supported by multinational corporations (TNCs) in 2004: Ponds Femina Miss Beautiful Skin, Miss Perfect

Ten, Miss Talented, Miss Athletic, SGI Miss Photogenic, HappyDent Miss Beautiful Smile, Sony My Miss India, Kidah Miss Sparkling Eyes, Miss Mirchi Listeners Choice, Ponds Googly Woogly Woosh Miss Congeniality, Miss Provogue 8888 and Richfeel Miss Beautiful Hair (*Times of India* 2005:2).
7  Phookan (2004) found that these TNCs sell their products in India: Avon, Burberrys, Calvin Klein, Cartier, Christian Dior, Estee Lauder, Elizabeth Arden, Lancome, Chambor, Coty, L'Oreal, Oriflame, Revlon, Yardley, Wella, Schwarzkopf, Escada, Nina Ricci, Rochas, Yves Saint Laurant, Shiseido, Unilever and Palmolive.

## Chapter 4   The Modern Housewife

1  Fewer than 3 per cent of the 408 families who participated in the survey questionnaire had a parent or grandparent in a retirement home.
2  The *Mahabharata* was written by the sage Bhagavan Vedavyasa (or Vyasa). The stories which comprise the *Mahabharata* are voluminous. The most popular modern version of the saga is by Rajagopalachari (1999), originally published in 1951 and reprinted 36 times. The stories which form the basis of the *Ramayana* were written by the sage Valmeeki, though it is uncertain whether he composed them or simply recorded them. As with the *Mahabharata,* the most popular modern version was written by Rajagoplachari (2000) in 1951. It has been reprinted 33 times. Both sagas were converted into television dramas in the 1990s and both achieved very high viewer ratings (Mankekar 1993 and 1999).
3  The universe was said to be created by the Goddess Parashakti or the God Shiva (male), or a combination of Shakti (parakruthi or female) and Shiva (purusha or male). Parashakti has had three incarnations. The first is Parvati, the beautiful yet dutiful wife of Shiva. The second reincarnation is Lakshmi, the wife of Vishnu. She is the Goddess of wealth, love, goodness. The third is Saraswathi, the wife of Brahma, the Goddess of knowledge, arts, music and literature.
4  Ferguson (1999:170) writes that African women were subjected to the same types of reforms. In the early 20th century women were trained in 'mothercraft' and 'housecraft'. 'In order to turn supposedly lazy and immoral urban African women into "home-proud housewives for their wage-earning husbands (Hansen 1997:444)", instruction was provided (begun by missionaries and later taken up by government service organisations and private companies) in sewing, hygiene, laundry, handicrafts, cooking, home decorating and sometimes reading and writing.'
5  The survey results in Trivandrum show that the percentage of wives with salaried work is about the same in Hindu and Christian families, 36 per cent and 38 per cent respectively.
6  The conflict involved physical violence. In a study of domestic violence in seven Indian cities in the period 1997–1999, Trivandrum ranked highest (INCLEN/ICRW 2000).
7  Waldrop 2001 found a similar division of labour in food preparation between household women and domestic servants in elite Delhi families.
8  Frøystad (2005) and Waldrop (2001) both found this same reticence to talk about differentiation of tasks in terms of caste among elite families in their ethnographies in two North Indian cities.

9 Food should not only give life to the living, but food is also regarded as essential for the soul's passage through the first stage of the journey between this life and the next. All of the members of the extended family gather at the home of the deceased and provide food for the journey. In the 19th and early 20th century, each family member placed three mouthfuls of 'mouth rice' (*wyakkuaruy*), three mouthfuls of water and three mouthfuls of *pusiman* leaves in the mouth of the deceased. The Hindu reformer Padmanabham is credited with changing this practice, which he viewed as wasteful and messy. Today only water and leaves are placed in the deceased's mouth.

10 Garnett's (2007) study of changing food refrigeration in the UK reveals that space was also relevant to changing in food storage practices. She points out that as late as 1970 only 60 per cent of the UK population had a refrigerator. Her research indicates that refrigerators were seen by many people as a replacement for food cellars and cool rooms as a place to store foods. She argues that new housing designs in the 70s and 80s, which had fewer spaces set aside for cabinets and shelves, were partly responsible for the interest in the refrigerators.

## Chapter 5   Exercising the Extended Family

1 By the end of the 1950s, Kerala had 25 full-time family planning clinics offering financial incentives for sterilisation of both men and women. Mass sterilisation was encouraged in 'sterilisation camps'. The camps were organised in the style of religious retreats. Lasting anywhere from two weeks to a month, the retreats featured parades, slogans, speeches and classes all glorifying the small family and attaching a sense of patriotism to sterilisation. They culminated in what one government report from 1977 referred to as 'massive' vasectomies and tubectomies of thousands of men and women. This report's 'dedication' was 'to 78,432 individuals, who motivated by responsibility to their families and love of their country, accepted sterilisation at two month-long camps at Ernakulam...' (Government of Kerala 1972). In 1976/77, 205,000 people were sterilised in Kerala (Jeffrey 1992:199).

2 In the early part of the 20th century, family members were expected to remain sequestered for up to two weeks; today the sequestration does not usually last more than three days.

3 Ayyappan is the son of the gods Shiva and Mohini, a female form of Vishnu.

4 *Nayapam* – rice, jaggery and cardamom, all fried in oil.
*Muryukka* – black gram powder, fried in oil.
*Achappam* – rice flour mixed with sugar and coconut milk, and shaped like a flower.
*Murithirikothu* – rice powder, and jaggery.
*Pakkavada* – spicy mix of fried chilli, gram powder and rice powder.

5 According to the *Kerala Economic Review* (2003), 93 per cent of all births in Kerala in 2003 took place at a medical facility.

6 During the sequestion, Meena was served special foods, bathed in herbs (*vethideel*) and massaged. Meenakshi told me that the purpose of this treatment was to nurse the mother back to strength and to wash away the vestiges of birth pollution.

7 Gardner makes a similar observation about dowry in Bangladesh where dowry is not regarded as the property of the bride. 'A wife's wedding gold, for instance is frequently sold by her in-laws: there is little she can do to prevent her property being absorbed by her husband's household (1995:179).'
8 Johnson (1998) found that the situation in two communities he studied in the Philippines, where men pay bride-wealth, mirrored the situation in Kerala concerning education and dowry. There, women with high levels of education could demand higher bride-wealth from the families of grooms.
9 Members of the 'scheduled castes' (the former lower tier of castes, including the 'untouchable' castes) are allocated quotas at universities and are provided stipends. Many Nair feel that this quota system is no longer fair. Members of the Ezhava caste are still eligible for university allocations and stipends even though as shown in Chapter 2, the economic strengths of families of the Nair and Ezhava are approaching parity.
10 Freeman writes about a family in Barbados, 'The young women in my survey continue to believe in the extended family... (even if) only half of these women are predominately supported by others and only a small proportion of these by husbands or partners (2000:130).'

## Chapter 6 Work Migration

1 According to Lall (2000), after the opening of India in 1992, new initiatives were taken by the Indian government to encourage NRI repatriation, including:
 - Repatriated money could be used in real estate development, construction of houses, financing of housing development and infrastructure such as roads and bridges, without special taxation. Houses and real estate could be bought and sold without having to seek prior approval of the government.
 - Repatriated income could be used to buy India Development Bonds which yielded 9.5 per cent interest.
 - NRIs who returned to India, either for employment or business were granted exemptions from declaring their foreign currency accounts and other forms of foreign currency assets, provided they had stayed abroad a minimum of two years. They were also allowed to retain foreign currency accounts without any limit on the balance.
 - Taxes on dividends from investments made in foreign exchange were reduced.
2 Cavita talked frequently about her good fortune, in contrast to that of her cousins who stayed and experienced the invasion. She said that the invasion came as a total surprise. When it started, the Iraqi army moved so fast that her cousins were only able take with them what they could put in their pockets. Public transportation was disabled so people had to walk 80 or 90 km to the Saudi border. From there they were taken by bus to Jordan, from where they flew home. Altogether the journey took 14 days.
3 During the Gulf War, 150,000 Indians were evacuated from the Gulf. In 1992, 109,000 Indians returned to Kuwait (Lall 2000).
4 Migration as a family enterprise applies to migration in many parts of Asia (see Gamburd 2000, and Gardner 1995). As Johnson wrote about Filipino

work migrants 'people did not work for themselves (*baran-baran nila*), but for their family, for their parents (1998:231)'.
5 Bourdieu defined habitus as 'Schemes of classification (which) owe their specific efficacy to the fact that they function below the level of consciousness and language, beyond the reach of introspective scrutiny or control by the will (1984:466).'
6 Caroline Osella, based on her work with Gulf migration from Malabar used the term 'double habitus' to capture the social reality of Gulf workers.

## Chapter 7 Material, Discursive and Performative Contributions to Consumption

1 Formerly, there was a caste called the *dobi* who washed clothes for the Nair and Brahmin families. Clothes were collected, boiled in large vats and then returned to their owners. As the system of caste-based work declined, the task of washing clothes was taken over by the women in the household, or given to a domestic servant.
2 Interview with Dr. Jacob Pulickan, Coordinator, Centre for Gandhian Studies, University of Kerala.
3 The astrological principles of *thachusasthram* are still followed in home construction, such as the building's orientation, the placement of rooms and the location of doors and windows. There are also auspicious dates for building and for occupying the home once it is finished. We were struck by the bad luck of Suba, Vishnukumar, and their three-year old son. Suba and Vishnukumar constructed a home and consulted an astrologer about auspicious dates for moving in. He gave two dates several months apart. When delays in construction caused them to miss the first date, they were forced to live with Suba's parents until the second auspicious date three months later, even though she was pregnant and the house was ready for occupation only a few weeks after the first date.
4 Brochmann describes a similar development in Sri Lanka. Work migrants changed village landscapes with their bigger houses, which were long-term projects. Many of them were never completed because 'the standard had obviously been too ambitious (1982:20).'
5 The disadvantage is that these smuggled air conditioners obviously do not come with a service contract, which I found Trivandrum families to be very concerned about.

## Chapter 8 Frictionless Political and Religious Ideologies

1 Chapter 2 takes up the unfavourable treaties to which the British subjected Cochin and Travancore and consequent resentment and resistance to the British in southwest Kerala.
2 This was the antithesis to the US post-World War II vision of development, in which poor countries were seen to develop through 'cultural emulation' of the United States (Foster 1965; Silvert 1977). In the 1960s, a critique of this concept of development emerged in Latin American. The 'Dependency Theorists' argued that US style development was not only failing to advance

Latin American economies and societies, but that it was also the source of 'underdevelopment' (Cardoso and Faletto 1979; Frank 1972). By the 1990s many North American and European intellectuals, among them Wolfgang Sachs (1992) and Arturo Escobar (1995) pronounced the 'end' of development as a project for global social and economic reforms. Others still argue for a change along the lines proposed by Gandhi, namely drawing on local knowledge. Ferguson, for example, argued that 'Much that was understood as backward and disappearing seems today to be most vital (1999:250).'

3 Khilnani (1998) claims that Indira Gandhi used defence of the planned economy to justify the declaration of a number of anti-democratic laws in the 1970s.
4 There had been a moderate step in the direction of liberalisation in 1985 when duties were lowered on capital goods and a number of industries were de-licensed (Lakha 1999).
5 Manmohan Singh would later become India's Prime Minister after the Congress Party won the majority of seats in parliament in the May election of 2004.
6 Using the words of Lury and Warde to describe how consumption was legitimised in post-World War II Europe.
7 The UNDP's Human Development Reports, issued each year, keep track of progress towards these indicators of human development around the world.
8 According to Franke and Chasin (1996), examples of those who have been excluded are fishing people, stone cutters, domestic servants, tribal peoples and migrants from Tamil Nadu.
9 The KSSP was awarded the 'alternative Nobel Prize' in 1996.
10 The interest in Hinduism at the end of the last millennium may be related to the increase in influence of the BJP and its Hindu nationalist agenda over the 1990s, though I did not find much wholehearted support for the BJP in Kerala in 2002.
11 According to Daniel (1984) participation among middle class men from neighbouring Tamil Nadu is also increasing.
12 Osella and Osella (2003) describe the two routes to *Sabarimala*, the long route being 65 km. long and taking several days to complete. The short route, taken by the majority of Trivandrum middle class males, is about 6 km. long, and can be traversed in about four hours. On arriving at the shrine's entrance, the pilgrims smash their coconuts against a rock. This represents the disintegration of the individual ego and the opening of self to the God.
13 This export and reimport of devotionalism is discussed in a work in progress by Kathinka Frøystad, 'U-turn Globalization: The Case of New Age in India', Department of Social Anthropology, University of Oslo.
14 This is similar to what Meyer (2002) found among Protestants in Ghana. There, Pentecostal missionaries have associated excessive Western consumption with a path that would end in 'hellfire'. Western goods are portrayed as 'dangerous' and 'enchanted', and must be counteracted by prayer.
15 There is a debate about the social significance of the land reforms. Jeffrey (1992) plays down the significance of the Land Reform Act, writing that the small parcels of land it awarded to tenants improved their lives, but did not significantly improve their economic conditions. Franke and Chasin (1996)

analysed redistribution of land holdings and concluded that they had a significant effect on class and caste inequality as well as on family income.

16  One of the posters displayed at the Centre for Science and Environment in New Delhi has a slogan which expresses the same sentiment this way: 'Frugality is an Indian tradition. Modernity sees frugality as poverty.'

## Chapter 9   Television: 'Everyone is Watching'

1  Access to both Al Jazeera and CNN gave our neighbours in Trivandrum a more balanced view of events on and after September 11, 2001 than most people in Europe and North America. People were hypnotised, as people everywhere were, by the unfolding of these events. Televisions in virtually every home were tuned to international news networks for days afterwards. The ethnographic roles were reversed, as I was sought out to watch with friends and families in the neighbourhood while they assessed my reactions.

2  To recapitulate the story (related in Chapter 4), Sita, the wife of the God-king Rama (an incarnation of Vishnu), is kidnapped by a foreign king and held captive. She is eventually rescued, but then her loyalty and chastity are doubted by Rama. She is forced to prove her innocence by withstanding a literal 'trial by fire'; she comes through unscathed.

3  According to Bullis (1997:89), television advertising billings in India grew 150 per cent between 1988 and 1992, and then increased another 50 per cent from 1992 to 1995.

## Chapter 10   Conclusion

1  Technology leapfrogging was at the heart of the work of the World Commission on the Environment (WCDE), the landmark UN study that resulted in the book *Our Common Future* (WCDE 1987) and a key concept in the textbook by the scientific contributors to the WCDE, *Energy for a Sustainable World* (Goldemberg et al. 1988).

2  For a discussion of the effects of changing dwelling size on energy use, see Wilhite and Norgard (2004). Per capita dwelling size in Scandanavia is influenced by the aging population and high incidence of divorce. In the Scandinavian cities of Oslo, Norway and Gothenburg, Sweden, more than 50 per cent of dwellings are inhabited by only one adult (Wilhite and Ling 1992).

# References

Abu-Lughod, L. 1999. The Interpretation of Culture(s) after Television. In Sherry Ortner (ed.), *The Fate of Culture: Geertz and Beyond*, pp. 110–135. Berkeley, Los Angeles, London: University of California Press.

Abu-Lughod, L. 1990. The Romance of Resistance: Tracing Transformations of Power Through Bedouin Women. In P. R. Sanday and R. G. Goodenough (eds), *Beyond the Second Sex: New Directions in the Anthropology of Gender*, pp. 311–338. Philadelphia: University of Pennsylvania Press.

Akrich, M. 2000. The Description of Technical Objects. In W. Bijker and J. Law (eds), *Shaping Technology/Building Society*, pp. 205–224. Cambridge Massachusetts: The MIT Press.

Alexander, W. 2004. Empowered women create high well-being: The case of Kerala within India. Conference paper, An International Dialog on Sustainable Development, Sustainable Development Institute, College of Menominee Nation, Keshena, Wisconsin, June.

Appadurai, A. 1996. *Modernity at Large: Cultural Dimensions of Globalization*. Minneapolis and London: University of Minnesota Press.

Appadurai, A. 1986. Introduction: Commodities and the Politics of Value. In A. Appadurai (ed.), *The Social Life of Things: Commodities in a Cultural Perspective*, pp. 3–63. Cambridge: Cambridge University Press.

Arcinas, F. R. 1986. The Philippines. In G. Gunatilleke (ed.), *Migration of Asian Workers to the Arab world*. Tokyo: United Nations University.

Argarwal, A., S. Narain and I. Khurana. 2001. *Making Water Everybody's Business*. New Delhi: Centre for Science and Environment.

Arunima, G. 1996. Multiple meanings: Changing conceptions of matrilineal kinship in nineteenth and twentieth century Malabar. *The Indian Economic and Social History Review*, 33(3): 283–307.

Askew, K. and R. Wilk (eds) 2002. *The Anthropology of the Media: A Reader*. Malden, Massachusetts and Oxford: Blackwell Publishers.

Babb, L. A. 1986. *Redemtive Encounters: Three Modern Styles in the Hindu Tradition*. Berkeley, Los Angeles, London: Berkeley University Press.

Bannerji, H. 1991. Fashioning a self: Educational proposals by and for women in popular magazines in Colonial Bengal. *Economic and Political Weekly*, 26(43): 50–62.

Barker, C. 2000. *Cultural Studies: Theory and Practice*. London, Thousand Oaks, New Delhi: Sage Publications.

Barley, N. 1986. *The Innocent Anthropologist: Notes from a Mud Hut*. London: Penguin.

Barthes, R. 1973. *Mythologies*. London: Paladin.

Baumann, Z. 1988. *Freedom*. Milton Keynes: Open University Press.

Benedict, R. 1946. *The Chrysanthemum and the Sword: Patterns of Japanese Culture*. Boston: Houghton Mifflin.

Bourdieu, P. 1998. *Practical Reason*. Cambridge: Polity Press.

Bourdieu, P. 1984. *Distinction: A Social Critique of the Judgment of Taste*. London: Routledge and Kegan Paul.

Bourdieu, P. 1977. *Outline of a Theory of Practice*. Cambridge: Cambridge University Press.
Brewer, J. and F. Trentmann. 2003. Introduction: Space, Time and Value in Consuming Cultures. In J. Brewer and F. Trentmann (eds), *Consuming Cultures, Global Perspectives: Historical Trajectories, Transnational Exchanges*, pp. 1–18. Oxford and New York: Berg.
Brochmann, G. 1982. Female migration from Sri Lanka to the Middle East. Prio Working Paper 6/86, International Peace Research Institute, Oslo, Norway.
Brown, B. 1983. The impact of male labour migration on women in Botswana. *African Affairs*, 82: 367–388.
Brown, S. and D. Turley. 1997. *Consumer Research: Postcards From the Edge*. London: Routledge.
Bullis, D. 1997. *Selling to India's Consumer Market*. Westport: Quorum Books.
Burke, T. 2003. *Lifebuoy Men, Lux Women: Commodification, Consumption and Cleanliness in Modern Zimbabwe*. Durham and London: Duke University Press.
Cardoso, F. and E. Faletto. 1979. *Dependency and Development in Latin America*. Berkeley: University of California Press.
Carrier, J. and D. Miller. 1999. From private virtue to public vice. In H. Moore (ed.), *Anthropological Theory of Today*. Cambridge and Oxford: Polity Press.
Census of India. 2001. www.censusindia.net.
Census of India. 1991. www.censusindia.net.
Chakravarti, Sudeep. 1995. The middle class: Hurt but hopeful. *India Today*, 12 April.
Chakravarty, Suhash. 2003. *The Raj Syndrome*. New Delhi: Oxford University Press.
Chanda, P. S. 1991. Birthing terrible beauties: Feminisms and 'Women's Magazines'. *Economic and Political Weekly*, 26(43): 67–70.
Chandhoke, N. 2005. 'Seeing' the State in India. *Economic and Political Weekly*, March 12: 1033–1039.
CI-ROAP. 1998. A Discerning Middle Class? A Preliminary Enquiry of Sustainable Consumption Trends in Selected countries in the Asia Pacific Region. Consumers International Regional Office for Asia and the Pacific (CI-ROAP), Penang, Malaysia.
Cohen, B. C. and R. Wilk with B. Stoeltje. 1996. Introduction. In B. C. Cohen, R. Wilk and B. Stoeltje (eds), *Beauty Queens on the Global Stage: Gender, Contests and Power*, pp. 1–11. New York and London: Routledge.
Colloredo-Mansfeld, R. 1999. *The Native Leisure Class: Consumption and Cultural Creativity in the Andes*. Chicago and London: The University of Chicago Press.
Comaroff, John and Jean Comaroff. 1992. *Ethnography and Historical Imagination*. Boulder: Westview Press.
Connell, R. W. 1987. *Gender and Power*. Cambridge: Polity Press.
Cooper, G. 1998. *Air Conditioning America: Engineers and the Controlled Environment, 1900–1960*. Baltimore: The John Hopkins University Press.
Corbridge, S. and J. Harriss. 2000. *Reinventing India: Liberalization, Hindu Nationalism and Popular Democracy*. Cambridge: Polity Press.
Counihan, C. 1999. *The Anthropology of Food and Body: Gender, Meaning and Power*. New York: Routledge.
Cowan, R. 1989. *The Ironies of Household Technology from the Open Hearth to the Microwave*. London: Free Association Books.

Daniel, E. V. 1984. *Fluid Signs: Being a Person the Tamil Way*. Berkeley: University of California Press.
Davidar, D. 2001. *The House of Blue Mangoes*. New Delhi: Penguin India.
Dempsey, C. G. 2001. *Kerala Christian Sainthood: Collisions of Culture and Worldview in South India*. Oxford and New York: Oxford University Press.
Department of Transport, India. 2002. Category-wise Growth of Motor Vehicles in Kerala. Delhi: Government of India.
Devi, L. K. R. 2002. Education, Employment, and Job Preference of Women in Kerala: A micro-level case study. Discussion Paper No. 42, Kerala Research Programme on Local Level Development, Centre for Development Studies, Thiruvananthapuram, Kerala.
Devika, J. 2003a. Women's History or History of En-Gendering?: Reflections on Gender and History-Writing in Kerala. Centre for Development Studies, Trivandrum.
Devika, J. 2003b. Beyond Kulina and Kulata: The Critique of Gender Difference in the Writings of K. Saraswati Amma. Centre for Development Studies, Trivandrum.
Devika, J. 2002a. Domesticating Malayalees: Family Planning, The Nation and Home-Centered Anxieties in Mid-20$^{th}$ Century Keralam. Working Paper Series 340. Trivandrum: Centre for Development Studies.
Devika, J. 2002b. Family Planning as 'Liberation': The Ambiguities of 'Emancipation from Biology' in Keralam. Working Paper Series 335. Trivandrum: Centre for Development Studies.
Douglas, M. and B. Isherwood. 1979. *The World of Goods: Towards an Anthropology of Consumption*. Harmondsworth: Penguin.
Dreze, J. and A. Sen (eds). 1996. *Indian Development: Selected Regional Perspectives*. New Delhi: Oxford University Press.
Dube, L. 1997. *Women and Kinship: Comparative Perspectives on Gender in South and South-East Asia*. Tokyo: The United Nations University Press.
Dumont, L. 1980. *Homo Hierarchicus: The Caste System and its Implications*. Chicago: The University of Chicago Press.
Eapen, M. and P. Kodoth. 2004. Discrimination Against Women in Kerala: Engaging Indicators and Processes of Well Being. Unpublished manuscript, Centre for Development Studies, Trivandrum.
Eapen, M. and P. Kodoth. 2001. Family Structure, Women's Education and Work: Re-examining the High Status of Women in Kerala. Paper published by the Centre for Development Studies, Trivandrum, Kerala.
*Economic Review*. 2000. Thiruvananthapuram: State Planning Board.
Engineer, A. A. 1995. *Kerala Muslims: A Historical Perspective*. Delhi: Ajanta Publications.
Epstein, T. S. 1962. *Economic Development and Social Change in South India*. Manchester: Manchester University Press.
Escobar, A. 1995. *Encountering Development: The Making and Unmaking of the Third World*. Princeton, NJ: Princeton University Press.
Fahey, S. 1999. Vietnam's Women in the Renovation Era. In K. Sen and M. Stivens (eds), *Gender and Power in Affluent Asia*. London and New York: Routledge.
Falk, P. 1997a. The Geneology of Advertising. In P. Sulkunen, J. Holmwook, H. Radner and G. Schulze (eds), *Constructing the New Consumer Society*. London: Macmillan Press Limited.

Falk, P. 1997b. The Benetton-Toscani Effect: Testing the Limits of Conventional Advertising. In M. Nava, A. Blake, I. MacRury and B. Richards (eds), *Buy This Book: Studies in Advertising and Consumption*, pp. 64–86. London and New York: Routledge.

Falk, R. A. 1999. *Predatory Globalization: A Critique*. Cambridge: Polity Press.

Featherstone, M. 1990. *Global Culture: Nationalism, Globalization and Modernity*. Newbury Park: Sage.

Ferguson, J. 1999. *Expectations of Modernity: Myths and Meanings of Urban Life on the Zambian Copperbelt*. Berkeley, Los Angeles, London: University of California Press.

Fernandes, L. 2000. Nationalizing 'the global': media images, cultural politics and the middle class in India. *Media, Culture and Society*, 22(5): 611–628.

Foster, G. M. 1965. Peasant society and the image of the limited good. *American Anthropologist*, 67: 293–315.

Frank, A. G. 1972. The Development of Underdevelopment. In J. Cockroft, A. K. Frank and D. Johnson (eds), *Dependence and Underdevelopment*. New York: Doubleday.

Franke, R. 2002. Caste, Class, and Mobility in a Kerala Village. Lecture at the Trivandrum Press Club, 27 May.

Franke, R. and Barbara H. Chasin. 1996. Is the Kerala Model Sustainable? Lessons from the Past. Paper presented at the International Conference on Kerala's Development Experience: National and Global Dimensions, December, New Delhi.

Freeman, C. 2000. *High Tech and High Heels in the Global Economy: Women, Work, and Pink Collar Identities in the Caribbean*. Durham: Duke University Press.

Frøystad, K. 2005. *Blended Boundaries: Caste, Class and Shifting Faces of 'Hinduness' in a North Indian City*. New Delhi: Oxford University Press.

Fuller, C. J. 1992. *The Camphor Flame: Popular Hinduism and Society in India*. Princeton and Oxford: Princeton University Press.

Fuller, C. J. 1991. Kerala Christians and the Caste System. In Dipankar Gupta (ed.), *Social Stratification*, pp. 195–212. Bombay, Calcutta and Madras: Oxford University Press.

Fuller, C. J. 1976. *The Nayars Today*. Cambridge: Cambridge University Press.

Gallagher, K. M. 1991. An Exploration into the causes of squatting in the Kathmandu Valley. Master's thesis, Tribhuvan University, Kathmandu.

Gamburd, M. R. 2000. *The Kitchen Spoon's Handle: Transnationalism and Sri Lanka's Migrant Housemaids*. Ithaca and London: Cornell University Press.

Gardner, K. 1995. *Global Migrants, Local Lives: Travel and Transformation in Rural Bangladesh*. Oxford: Clarendon Press.

Garnett, T. 2007. Food refrigeration: What is the contribution to greenhouse gas emissions and how might emissions be reduced? A working paper produced as part of the Food Climate Research Network, Centre for Environmental Strategy, University of Surrey, England.

Garon, S. 2003. Japan's post-war 'consumer revolution', or striking a 'balance' between consumption and savings. In J. Brewer and F. Trentmann (eds), *Consuming Cultures, Global Perspectives: Historical Trajectories, Transnational Exchanges*, pp. 189–218. Oxford and New York: Berg.

Giddens, A. 1979. *Central Problems in Social Theory: Action, Structure and Contradiction in Social Analysis*. Berkeley: University of California Press.

Ginsburg, F., L. Abu-Lughod and B. Larkin. 2002. *Media Worlds: Anthropology on New Terrain*. Berkeley, Los Angeles and London: University of California Press.

Gledhill, C. 1997. Genre and Gender: The Case of Soap Opera. In S. Hall (ed.), *Cultural Representation and Signifying Practices*, pp. 365–386. London, Thousand Oaks and New Delhi: Sage Publications.

Goldemberg, J., Johansson, T. B., Reddy, A. K. N. and R. H. Williams. 1988. *Energy for a Sustainable World*. New Delhi: Wiley Eastern Limited.

Goody, J. and I. Watt. 1968. The Consequences of Literacy. In J. Goody (ed.), *Literacy in Traditional Societies*, pp. 27–69. Cambridge: Cambridge University Press.

Gopikuttan, G. 1990. House construction boom in Kerala. *Economic and Political Weekly*, 15: 2083–2088.

Gough, K. 1962a. Nayar: Central Kerala. In D. Schneider and K. Gough (eds), *Matrilineal Kinship*. Berkeley and Los Angeles: University of California Press.

Gough, K. 1962b. Modern Disintegration of Matrilineal Decent Groups. In D. Schneider and K. Gough (eds), *Matrilineal Kinship*. Berkeley and Los Angeles: University of California Press.

Government of India. 2005. Economic Survey, Review of Developments: Consumption, savings and investment. *Union Budget and Economic Survey*. Delhi: Ministry of Finance.

Government of Kerala. 1972. Family planning in Kerala. Trivandrum: Kerala Government Publications.

Gulati, L. 2003. Full circle: Women's studies sans institutions. In D. Jain and P. Rajput (eds), *Narratives from the Women's Studies Family*. New Delhi: Sage Publications.

Gupta, D. 2000. *Mistaken Modernity: India Between Two Worlds*. New Delhi: HarperCollins.

Hall, S. 1973. Encoding and Decoding in the TV Discourse, reprinted in S. Hall, I. Connell and L. Cutti (1981) (eds), *Culture, Media, Language*. London: Hutchinson.

Halliburton, M. 1998. Suicide: A paradox of development in Kerala. *Economic and Political Weekly*, 33(36 and 37): 2341–2345.

Hansen, K. T. 2000. *Salalula: The World of Second Hand Clothing and Zambia*. Chicago and London: The University of Chicago Press.

Hansen, K. T. 1997. *Keeping House in Lusaka*. New York: Colombia University Press.

Harilal, K. N. and M. Andrews. 2000. Building and builders in Kerala: Commodification of buildings and labour market dynamics. Discussion Paper 22, Keral Research Programme on Local Level Development, Centre for Development Studies, Thiruvananthapuram, Kerala.

Harilal, K. N. 1986. Kerala's Building Industry in Transition: A Study of the Organisation of Production and Labour Process. M. Phil. Dissertation, Centre for Development Studies, Thiruvananthapuram.

Hart, K. 1973. Informal income opportunities and urban employment in Ghana. *Journal of Modern African Studies*, 11(1): 61–79.

Holston, J. 1999. Spaces of Insurgent Citizenship. In James Holston (ed.), *Cities and Citizenship*. Durum and London: Duke University Press.

Hochschild, A. R. (with Anne Machung). 2003. *The Second Shift*. London: Penguin.

Hooper, B. 1998. 'Flower Vase and Housewife': Women and Consumerism in post-Mao China. In K. Sen and M. Stivens (eds), *Gender and Power in Affluent Asia*. London and New York: Routledge.
INCLEN/ICRW. 2000. Indiasafe: Studies of abuse in the family environment in India – A summary report. Internation clinical epidemiologists network, India.
India Census. 1991. Government of India, New Delhi.
*India Today*. 2000. 1975–2000, 25 incredible years. December 25: 48.
Issac, T. (with R. W. Franke). 2000. *Local Democracy and Development: People's Campaign for Decentralized Planning in Kerala*. New Delhi: Left World Books.
Iyer, L. K. A. 1981. *The Tribes and Castes of Cochin*. New Delhi: Cosmo publications.
Jackson, R. T. 1990. The cheque's in the mail: The distribution of dependence on overseas sources of income in the Philippines. *Singapore Journal of Tropical Geography*, II(2): 75–86.
Jain, P. and Mahan, R. 1996. Introduction. In Jain, P. and R. Mahan (eds), *Women Images*, pp. 11–34. Jaipur and New Delhi: Rawat Publications.
Jeffrey, R. 1992. *Politics, Women and Well-Being: How Kerala Became a Model*. New Delhi: Oxford University Press.
Jeffrey, R. 1978. Matriliny, Marxism and the birth of the Communist Party in Kerala, 1930–1940. *Journal of Asian Studies*, 38(1): 77–98.
Jhally, S. 2002. Image-based culture. In K. Askew and R. Wilk (eds), *The Anthropology of Media: A Reader*, pp. 327–336. Malden and Oxford: Blackwell Publishers.
Johnson, M. 1998. At Home and Abroad: Inalienable Wealth, Personal Consumption and Formulations of Femininity in the Southern Philippines. In D. Miller (ed.), *Material Culture: Why Some Things Matter*, pp. 215–238. London: UCL Press Ltd.
Joseph, A. 2002. Fools Paradise? In A. Narain (ed.), *Where the Rain is Born: Writing about Kerala*, pp. 87–96. New Delhi: Penguin Books.
Kariyil, A. 2000. *Church and Society in Kerala: A Sociological Study*. New Delhi: Intercultural Publications.
Kearney, M. 1995. The local and the global: the anthropology of globalization and transnationalism. *Annual Review of Anthropology*, 24: 547–565.
Kempton, W. and L. Lutzenhiser. 1992. Introduction. *Energy and Buildings*, 18: 171–176.
*Kerala Economic Review*. 2003. Kerala State Planning Board, Government of Kerala, Thiruvananthapuram.
Khilnani, S. 1998. *The Idea of India*. London: Penguin Books.
Kolenda, P. 1991. The Ideology of Purity. In D. Gupta (ed.), *Social Stratification*, pp. 28–34. Delhi: Oxford University Press.
Kolshus, T. S. 2005. Myten om moderniteten – et litt for polemisk essay med et litt for alvorlig budskap. *Norsk Antropologisk tidsskrift*, 1(16): 34–48.
Kopytoff, I. 1986. The Cultural Biography of Things: Commoditization as Process. In A. Appadurai (ed.), *The Social Life of Things: Commodities in a Cultural Perspective*. Cambridge: Cambridge University Press.
Kulkarni, D. 2001. The renunciative way of life. *The Hindu*. December 10.
Kurian, V. 2002. 'Hand wash' campaign in Kerala raises a stink. *The Hindu Business Line*, November 6.
Lakha, S. 1999. The State, Globalisation and Indian Middle-class Identity. In M. Pinches (ed.), *Culture and Privilege in Capitalist Asia*. London and New York: Routledge.

Lall, M. C. 2000. *India's Missed Opportunity: India's Relationship with the Non Resident Aliens*. Aldershot: Ashgate.

Lash, S. and Urry, J. 1994. *Economies of Signs and Space*. London: Sage.

Latour, B. 2000. Where are the Missing Masses? The Sociology of a Few Mundane Artifacts. In W. Bijker and J. Law (eds), *Shaping Technology/Building Society*, pp. 225–258. Cambridge Massachusetts: The MIT Press.

Latour, B. 1993. *We Have Never Been Modern*. New York: Harvester Wheatsheaf.

Lawson, Tony. 1997. *Economics and Reality*. London and New York: Routledge.

Lee, M. 1998. *Women's Education, Work and Marriage in Korea: Women's Live's Under Institutional Conflicts*. Seoul: Seoul National University Press.

Liechty, M. 2003. *Suitably Modern: Making Middle-Class Culture in a New Consumer Society*. Princeton and Oxford: Princeton University Press.

Lindberg, A. 2001. Experience and Identity: A Historical Account of Class, Caste, and Gender Among the Cashew Workers of Kerala, 1930–2000. PhD Dissertation. Lund: Historiska institutionen vid Lunds universitet.

Lury, C. and A. Warde. 1997. Investments in the Imaginary Consumer: Conjectures Regarding Power, Knowledge and Advertising. In M. Nava, A. Blake, I. MacRury and B. Richards (eds), *Buy This Book: Studies in Advertising and Consumption*, pp. 87–102. London and New York: Routledge.

MacCannell, D. and J. MacCannell. 1993. Social Class in Post-modernity: Simulacram or Return of the Real? In C. Rojek and B. Turner (eds), *Forget Baudrillard*. London and New York: Routledge.

MacKinnon, C. 1989. *Towards a Feminist Theory of the State*. Harvard: Harvard University Press.

Mankekar, P. 1999. *Screening Culture, Viewing Politics: An Ethnography of Television, Womanhood, and Nation in Postcolonial India*. Durham: Duke University Press.

Mankekar, P. 1993. Television tales and a woman's rage: A nationalist recasting of Draupadi 'disrobing'. *Public Culture*, 5(3): 127–176.

Marcuse, H. 1964. *One-Dimensional Man*. Boston: Beacon Press.

Marriott, M. and R. B. Inden. 1977. Towards a Ethnosociology of the South Asian Caste System. In K. A. David (ed.), *The New Wind: Changing Identities in South India*.

Mazzarella, W. 2003. *Shoveling Smoke: Advertising and Globalization in Contemporary India*. Durham and London: Duke University Press.

Menon, A. S. 1998. *A Survey of Kerala History*. Madras: S. Viswanathan.

Meyer, B. 2002. Commodities and the Power of Prayer: Pentecostalist Attitudes towards Consumption in Contemporary Ghana. In J. Xavier and R. Rosaldo (eds), *The Anthropology of Globalization*, pp. 247–267. Malden, Mass: Blackwell Publications Ltd.

Miller, D. 2001a. *The Dialectics of Shopping*. Chicago and London: The University of Chicago Press.

Miller, D. 2001b. Introduction. In D. Miller (ed.), *Car Cultures*, pp. 1–34. Oxford and New York: Berg.

Miller, D. 1998. Introduction. In D. Miller (ed.), *Material Culture: Why Some Things Matter*, pp. 3–21. London: UCL Press Ltd.

Miller, D. 1995a. Consumption Studies as the Transformation of Anthropology. In D. Miller (ed.), *Acknowledging Consumption: A Review of New Studies*, pp. 264–295. London: Routledge.

Miller, D. (ed.) 1995b. *World's Apart: Modernity Through the Prism of the Local*. London: Routledge.

Miller, D. 1995c. Consumption As the Vanguard of History. In D. Miller (ed.), *Acknowledging Consumption: A Review of New Studies*, pp. 1–57. London: Routledge.

Miller, D. 1994. *Modernity: An Ethnographic Approach: Dualism and Mass Consumption in Trinidad*. Oxford & NY: Berg.

Miller, D. and D. Slater. 2000. *The Internet: An Ethnographic Approach*. Oxford: Berg.

Mills, M. B. 2003. *Thai Women in the Global Labor Force: Consuming Desires, Contested Selves*. New Brunswick, New Jersey and London: Rutgers University Press.

Moore, H. 1999. *Anthropological Theory Today*. Cambridge: Polity Press.

Moore, H. 1994. *A Passion for Difference*. Cambridge and Oxford: Polity Press.

Morley, D. 1995. Theories of Consumption in Media Studies. In D. Miller (ed.), *Acknowledging Consumption: A Review of New Studies*. London: Routledge.

Mukhopadhyay, C. C. and S. Seymour. 1994. Introduction and Theoretical Overview. In Carol C. Mukhopadhyay and Susan Seymour (eds), *Women, Education and Family Structure in India*. Boulder, San Francisco and Oxford: Westview Press.

Murickah, S. J. J. 2002. *Marriage and Human Fulfilment*. New Delhi: Classical Pub. Co.

Murray, C. 1981. *Families Divided: the Impact of Migrant Labour in Lesotho*. Cambridge: Cambridge University Press.

Narayan, S. (ed). 1968. *The Selected Works of Mahatma Gandhi*. Ahmedabad: Navajivan Publishing House.

Nash, J. A. 1998. On the Subversive Virtue: Frugality. In D. A. Crocker and T. Linden (eds), *The Ethics of Consumption: The Good Life, Justice, and Global Stewardship*, pp. 416–446. Lanham, Boulder, New York, Oxford: Rowman & Littlefield.

National Sample Survey Organisation. 2001/2. *Household Consumer Expenditure and Employment Situation in India, 2001–2002*. Delhi: Ministry of Statistics and Programme Implementation, Government of India.

Nava, M. 1997. Framing Advertising: Cultural Analysis and the Incrimination of Visual Texts. In M. Nava, A. Blake, I. MacRury and B. Richards (eds), *Buy This Book: Studies in Advertising and Consumption*, pp. 34–51. London and New York: Routledge.

O'Dougherty, M. 2002. *Consumption Intensified: The Politics of Middle-Class Daily Life in Brazil*. Durham and London: Duke University Press.

OECD. 2004. *World Energy Outlook*. Paris: OECD.

Olwig, K. F. 2003. Global Places and Place-identities. In Thomas H. Eriksen (ed.), *Globalisation: Studies in Anthropology*. London and Sterling Virginia: Pluto Press.

Ortner, S. B. 1999. Thick Resistance: Death and the Cultural Construction of Agency in Himalaya Mountaineering. In S. B. Ortner (ed.), *The Fate of 'Culture': Geertz and Beyond*, pp. 136–165. Berkeley: University of California Press.

Ortner, S. B. 1995. Resistance and the problem of ethnographic refusal. *Comparative Studies in Society and History*, 37(1): 173–193.

Osella, F. and C. Osella. 2003. 'Ayyappan saranam': Masculinity and the Sabarimala pilgrimage in Kerala. *Journal of the Royal Anthropological Institute*, 9: 729–754.

Osella, F. and C. Osella. 2000. *Social Mobility in Kerala: Modernity and Identity in Conflict*. London: Pluto Press.

Osella, F. and C. Osella. 1996. Articulation of physical and social bodies in Kerala. *Contributions to Indian Sociology*, 30(1): 37–68.

Packard, V. 1957. *The Hidden Persuaders*. London: Longmans.

Panikkar, K. N. 1989. *Against the Lord and State: Religion and Peasant Uprisings in Malabar, 1836–1921*. New Delhi: Oxford University Press.

Parayil, G. 2000. *Kerala: The Development Experience: Reflections on Sustainability and Replicability*. London and New York: Zed Books.

Peltzer, C. 1993. Socio-cultural Dimensions of Renovation in Vietnam: Doi Moi as Dialogue and Transformation in Gender Relations. In W. S. Turley and M. Seldon (eds), *Reinventing Vietnamese Socialism: Doin Moi in Comparative Perspective*. Boulder: Westview Press.

Phookan, M. 2004. Consumption and cosmetics, industry sector analysis. Report prepared for the Cosmetics and Personal Care Industry, Canada. Available on the web at www.strategis.ic.gc.ca/epic/internet/inimr-ri.nsf/en/gr116897e.html.

Pinches, M. 1999. Cultural Relations, Class and the New Rich of Asia. In M. Pinches (ed.), *Culture and Privilege in Capitalist Asia*. London and New York: Routledge.

Polanyi, K. 1957. *The Great Transformation: The Political and Economic Origins of Our Time*. Boston: Beacon Press.

Prakesh, G. 2003. Religion and Nation in Colonial India. In B. Meyer and P. Pels (eds), *Magic and Modernity: Interfaces of Revelation and Concealment*. Stanford: Stanford University Press.

Puthenkalam, S. J. 1977. Marriage and the Family in Kerala. *Journal of Comparative Family Studies Monograph Series*, pp. 127–166. Calgary: The University of Calgary.

Raheja, G. G. and A. G. Gold. 1994. *Listen to the Heron's Words: Reimagining Gender and Kinship in North India*. Berkeley: University of California Press.

Raj, K. N. and M. Tharakan. 1981. Agrarian Reform in Kerala and its Impact on Rural Economy – A Preliminary Assessment. In K. N. Raj and M. Tharakan (eds), *Agrarian Reform in Kerala and its Impact on the Rural Economy*, pp. 31–51. Geneva: International Labour Organisation.

Rajagopalachari, C. 2000. *Ramayana*. Mumbai: Bharatiya Vidy Bhavan.

Rajagopalachari, C. 1999. *Mahabharata*. Mumbai: Bharatiya Vidy Bhavan.

Rajashree, A. 2001. Advertisers feature. *The Hindu*, November 26, 2001.

Rip, A. and R. Kemp. 1998. Technological Change. In Steve Rayner and Elizabeth Malone (eds), *Human Choice and Climate Change*, pp. 327–400. Ohio: Battelle Press.

Robinson, C. 1999. *Tradition and Liberation: The Hindu Tradition in the Indian Woman's Movement*. Surrey: Curzon Press.

Rojek, C. 2004. The consumerist syndrome in contemporary society: An interveiw with Zygmunt Bauman. *Journal of Consumer Culture*, 4(3): 291–312.

Rowchowdry, A. 2001. Globalisation. *The Hindu*, November 21.

Sachs, W. (ed.). 1992. *The Development Dictionary: A Guide to Knowledge as Power*. London: Zed Books.

Saradamoni, K. 1999. *Matriliny Transformed: Family, Law and Ideology in Twentieth Century Travancore*. New Delhi, Walnut Creek and London: Sage Publications and Altamira Press.

Schama, S. 1987. *The Embarassment of Riches: An Interpretation of Dutch Culture in the Golden Age*. London: Collins.

Seabrook, J. 1992. The reconquest of India: the victory of international monetary fundamentalism. *Race and Class*, 34(1): 1–16.

Sen, C. T. 2004. *Food Culture in India*. Westport, Connecticut and London: Greenwood Press.

Sen, K. and M. Stivens (eds). 1998. *Gender and Power in Affluent Asia*. London and New York: Routledge.

Sharma, A. and A. Roychowdhury. 1996. *Slow Murder: The Deadly Story of Vehicular Pollution in India*. Delhi: Centre for Science and Environment.

Shiva, V. 2002. Saving lives or destroying lives? World Bank sells synthetic soap & cleanliness to Kerala: the land of health and hygiene. www.mindfully.org.

Shove, E. 2003. *Comfort, Cleanliness + Convenience: The Social Organization of Normality*. Oxford, New York: Berg.

Sigaut, F. 1994. Technology. In Tim Ingold (ed.), *Companion Encyclopedia of Anthropology*. London and New York: Routledge.

Silvert, K. H. 1977. *Essays in Understanding Latin America*. Philadelphia: Institute for the Study of Human Issues.

Simpson, B. 1998. *Changing Families: An Ethnographic Approach to Divorce and Separation*. Oxford and New York: Berg.

Slater, D. 1997. *Consumer Culture and Modernity*. Cambridge: Polity Press.

Soman, C. R. and V. R. Kuty. 2004. Population Registry of Life Style Diseases. Report by Health Action by People, Trivandrum.

Southerton, D., H. Chappels and B. Van Vliet (eds). 2004. *Sustainable Consumption: The Implications of Changing Infrastructures of Provision*. Cheltenham and Northampton: Edward Elgar.

Southerton, D., A. Warde and M. Hand. 2004. The Limited Autonomy of the Consumer: Implications for Sustainable Consumption. In D. Southerton, H. Chappells and B. Van Vliet (eds), *Sustainable Consumption: The Implications of Changing Infrastructures of Provision*. Cheltenham and Northampton: Edward Elgar Publishing, pp. 32–48.

South India Human Development Report. 2001. The National Council of Applied Economic Research, India. New Delhi: Oxford University Press.

Spaargaren, G. 2004. Sustainable Consumption: A Theoretical and Environmental Policy Perspective. In D. Southerton, H. Chappells and B. Van Vliet (eds), *Sustainable Consumption: The Implications of Changing Infrastructures of Provision*, pp. 15–32. Cheltenham and Northampton: Edward Elgar Publishing.

Spitulnik, D. 1993. Anthropology and mass media. *Annual Review of Anthropology* 22: 293–315.

Srinivas, M. N. 1996. *Village, Caste, Gender and Method: Essays in Indian Social Anthropology*. New Delhi: Oxford University Press.

Statistical Abstracts India. 2003. Ministry of Statistics and Programmes. Government of India, Delhi.

Stivens, M. 1998. Theorising Gender, Power and Modernity in Affluent Asia. In K. Sen and M. Stivens (eds), *Gender and Power in Affluent Asia*, pp. 1–34. London and New York: Routledge.

Stoler, A. L. 1995. *Race and the Education of Desire: Foucault's History of Sexuality and the Colonial Order of Things*. Durham and London: Duke University Press.

Story, J. 1999. *Cultural Consumption and Everyday Life*. London, Sydney, Auckland: Arnold Press.
Surendran, P. 1999. *The Kerala Economy: Development, Problems and Prospects*. Delhi: Vrinda Publications Ltd.
Tarlo, E. 1996. *Clothing Matters: Dress and Identity in India*. London: Hurst & Company.
The Art of Living. 2003. Broshure entitled 'Basic Course: The Healing Breath Technique'. Thiruvananthapurum: Vyakti Vikas Kendra.
*The Economist* (online). 1994. The poor get richer: India, consumption boom, Vol. 333, November 5.
*The Financial Express*. 2002. Public-Private Alliance to Promote Hygiene Habits in Kerala. Monday, April 15.
*The Hindu*. 2001b. Plumping for Fitness, December 6.
*Times of India*. 2005. Sizzling Performance, March 27.
Tomlinson, J. 1991. *Cultural Imperialism*. London: Routledge.
Törnquist, O. 1995. *The Next Left? Democratisation and Attempt to Renew the Radical Political Development Project – The Case of Kerala* (with P. K. Michael Tharakan). Uppsala: Nordic Institute of Asian Studies, NIAS Report Series, No. 24.
Tsing, A. 2002. Conclusion: the Global Situation. In J. X. Inda and R. Rosaldo (eds), *The Anthropology of Globalisation*, pp. 453–485. London: Blackwell Publishers.
Usha, V. T. 2004. Gender, Value, and Signification: Women and television in Kerala. Discussion paper No. 67, Derala Research Programmme on Local Level Development. Trivandrum: Centre for Development Studies.
Varma, P. K. 1998. *The Great Indian Middle Class*. New Delhi: Viking.
Vijayakumar, K. and S. Chattopadhyay. 1999. Energy Consumption Pattern: A Comparative Study. Report by the Centre for Earth Science Studies. Trivananthapuram.
Vinikas, V. 1989. Lustrum of the Cleanliness Institute, 1927–1932. *Journal of Social History*, 22(4): 613–629.
Visvanathan, S. 2002. The Status of Christian Women in Kerala. In A. Sharma (ed.), *Women in Indian Religions*, pp. 189–200. Oxford and New York: Oxford University Press.
Visvanathan, S. 1989. Marriage, birth and death: property rights and domestic relationship of Orthodox/Jacobite Syrian Christians of Kerala. *Economic and Political Weekly*, 24(24): 1341–1346.
Wadley, S. 1995. Women and the Hindu Tradition. In D. Jacobson and S. Wadley (eds) *Women in India: Two Perspectives*. Colombia, MO: South Asia Publications.
Walby, S. 1992. Post-Post-Modernism. In M. Barrett and A. Phillips (eds), *Destabilizing Theory*. Cambridge: Polity Press.
Waldrop, A. K. 2001. A Room With One's Own: Educated Elite People in New Delhi and Relations of Class. Doctoral dissertation, Department of Social Anthropology, University of Oslo.
Wallerstein, I. 1974. *The Modern World-System, Vol. 1*. New York: Academic Press.
Warde, A. 1997. *Consumption, Food and Taste*. London: Sage.

Warde, A. 1996. Afterword: The future of the sociology of consumption. In S. Edgell, K. Hetherington and A. Warde (eds), *Consumption Matters*. Oxford: Blackwell Publishers.

Werbner, P. 1990. *The Migration Process: Capital, Gifts and Offerings among British Pakistanis*. New York, Oxford and Munich: Berg.

Wilhite, H. 2007. Will efficient technologies save the world? A call for new thinking on the ways that end-use technologies affect energy using practices. *Proceedings from the ECEEE 2007 Summer Study*, pp. 23–30. Paris: Euopean Council for an Energy Efficient Economy.

Wilhite, H. 2005. Why energy needs anthropology. *Anthropology Today*, 21(3): 1–3.

Wilhite, H. 2002. An Assessment of the Energy Management Centre's Energy Clinic and Thaapabharani Programs. A Report prepared for the Energy Management Centre, Government of Kerala, 12 February.

Wilhite, H. and J. Norgard. 2004. Equating efficiency with reduction: A self-deception in energy policy. *Energy and Environment*, 15(6): 991–1010.

Wilhite, H., H. Nakagami, T. Masuda, Y. Yamaga and H. Haneda. 2001. A Cross-cultural Analysis of Household Energy-use Behavior in Japan and Norway. In D. Miller (ed.), *Consumption: Critical Concepts in the Social Sciences*, Vol. 4, pp. 159–177. London and New York: Routledge.

Wilhite, H., E. Shove, L. Lutzenhiser and W. Kempton. 2000. The Legacy of Twenty Years of Demand Side Management: We Know More about Individual Behavior But Next to Nothing About Demand. In E. Jochem, J. Stathaye and D. Bouille (eds), *Society, Behaviour and Climate Change Mitigation*, pp. 109–127. Dordrect: Kluwer Academic Press.

Wilhite, H. and L. Lutzenhiser. 1998. Social loading and sustainable consumption. *Advances in Consumer Research*, 26: 281–287.

Wilhite, H., H. Nakagami and C. Murakoshi. 1997. Changing Patterns of Air Conditioning Consumption in Japan. In P. Bertholdi, A. Ricci and B. Wajer (eds), *Energy Efficiency in Household Appliances*, pp. 149–158. Berlin: Springer.

Wilhite, H. and R. Ling. 1992. The person behind the meter: An ethnographic analysis of residential energy consumption in Oslo, Norway. *Proceedings from the ACEEE 1992 Summer Study on Energy Efficiency in Buildings*, 10: 177–186. Washington, D.C.: American Council For An Energy Efficient Economy.

Wilhite, H. and R. Wilk. 1987. A method for self-recording household energy use. *Energy and Buildings*, 10(1): 73–79.

Wilk, R. 2002a. Television, Time, and the National Imaginary in Belize. In F. D. Ginsburg, Lila Abu-Lughod and Brian Larkin (eds), *Media Worlds: Anthropology on New Terrain*, pp. 171–188. Berkeley and Los Angeles: University of California Press.

Wilk, R. 2002b. Culture and Energy Consumption. In Robert Bent, Lloyd Orr, and Randall Baker (eds), *Energy: Science, Policy and the Pursuit of Sustainability*, pp. 109–130. Washington D.C.: Island Press.

Wilk, R. 2001. Consumer Goods as Dialogue about Development. In D. Miller (ed.), *Consumption: Critical Concepts in the Social Sciences*, Vol. 2, pp. 34–54. London and New York: Routledge.

Wilk, R. 1999. Towards a useful multigenic theory of consumption. Proceedings of the 1999 ECEEE Summer Study. Paris: European Council for an Energy Efficient Economy.

Wilk, R. 1996. Connections and Contradictions: From the Crooked Tree Cashew Queen to Miss World Belize. In In B. C. Cohen, R. Wilk and B. Stoeltje (eds), *Beauty Queens on the Global Stage: Gender, Contests and Power*, pp. 217–233. New York and London: Routledge.

Wilk, R. 1989. Houses As Consumer Goods. In H. Rutz and B. Orlove (eds), *The Social Economy of Consumption*, pp. 101–120. Lanham: University Press of America.

World Commission on Environment and Development (WCED). 1987. *Our Common Future*. Oxford: Oxford University Press.

Wujastyk, D. 1998. *The Roots of Ayurveda*. Delhi: Penguin.

Younger, P. 2002. *Playing Host to the Deity: Festival Religion in the South Indian Tradition*. Oxford: Oxford University Press.

Zachariah, K. C. and S. I. Rajan. 2004. Gulf revisited: Economic consequences of emigration from Kerala. Working Paper 363, Centre for Development Studies, Thiruvananthapurum, Kerala.

Zachariah, K. C. and K. P. Kannan. 2002. Introduction. In K. C. Zachariah, K. P. Kannan and S. Irudaya Rajan (eds), *Kerala's Gulf Connection: CDS Studies on International Labour Migration from Kerala State in India*, pp. 1–13. Thiruvananthapurum: Centre for Development Studies.

Zachariah, K. C., E. T. Mathew and S. Irudaya Rajan. 2002a. Migrant Patterns and their Socio-economics. In K. C. Zachariah, K. P. Kannan and S. Irudaya Rajan (eds), *Kerala's Gulf Connection: CDS Studies on International Labour Migration from Kerala State in India*, pp. 13–47. Thiruvananthapurum: Centre for Development Studies.

Zachariah, K. C., E. T. Mathew and S. Irudaya Rajan. 2002b. Consequences of Migration: Soc-economic and Demographic Dimensions. In K. C. Zachariah, K. P. Kannan and S. Irudaya Rajan (eds), *Kerala's Gulf Connection: CDS Studies on International Labour Migration from Kerala State in India*, pp. 47–93. Thiruvananthapurum: Centre for Development Studies.

# Index

Page numbers in *italics* refer to tables and plates.

Abu-Lughod, L.   46, 164, 171
advertising
  air conditioners   *112*
  fairness (skin-lightening) products
    38, 153–154
  soaps   110
  and cleaning agents   126, 163
  and sponsorship   45–46, 150
  transnational agencies   166
  *see also under* television
*ahimsa* (truth and non-violence)
    129–130
air conditioners
  alternatives to   118–119, 175–176
  normalising   119–121, 126
  sales and changes in the political
    economy   117–118
  television advertising   160–161
  usage   118
  work migration   *101*, 118
Ambassador (cars)   123, 124
Amrityanandamayi (Amma),
    Sathguru Shri Matha   142
Amway   111
Appadurai, A.   89, 124, 153
'Art of Living' devotional movement
    143
astrology
  house design   113
  marriage compatibility   81, 82
ayurvedic medicine   106–107
  food beliefs   64
  soap products   111

Baba, Sathya Sai   142–144
Babb, L. A.   142, 143
'Baker house'   118–119
Baker, L.   118–119
baptism   73
Barley, N.   170–171
Bauman, Z.   124–125
beauty   38–40
  Western ideals of   42–48

beauty contests   42–43, 44–46
beauty products *see* cosmetics
beauty queens   42, 43, 160
Bharatiya Janata Party (BJP)   133
*bindi*   47
bodily cleanliness *see* cleanliness;
    soap
bodily shape of women   43–44
Bourdieu, P.   2, 6, 68, 102, 170
Brahmin caste, Nambuthiri   23, 24
Brewer, J. and Trentman, F.   167
British colonialism
  consumption as freedom from
    repressions of   134–135, 171
  house construction and design
    113
  soaps and cleanliness   110
  thriftiness and frugality   130
  and Western materialism   18

cars   121–125, 127
  advertising   163
  caste differences in ownership   28
  as dowry   84, 85
  migrant workers' ownership   94,
    101–102
    Gulf vs non-Gulf families   98,
      *99*, *100*
  sales   122
  and social status   122–123, 124,
    125, 127
cash dowries   84, 85, 86–87
caste   16–17
  attitudes to foreign goods
    134–136
  dowry practices   32–33, 79–80
  history   22–25
  house construction   113
  mobility and emulation, decline in
    importance of   28–29
  purity issues   62, 105
  reforms and blurring of caste
    hierarchy   22–29

same- and cross-caste marriage  32
socio-economic parity  25–28
*see also* Ezhava caste; Nair caste
Chakravarty, S.  56–57
Chandhoke, N.  131
child-directed advertising  162–163
childbirth  78
childhood rituals  73–74
*chorunu* (first feeding ritual)  73, *74*
Christians
  dowry practices  80, 85
  early history  17, 22–23
  family rituals  73–74
  religiosity  144–145
  work migration  90
*churidar-kurta* clothing  37
CI-ROAP (Consumer International)  9, 146
cleanliness  104–106, 126
  *see also* soap
'close distance' concept  47, 136, 160
clothes  *50, 59*
  men's  37–38, 129
  washing  52, 60–61, 105, 106
  women's  35–37, *46*
CNN  151–152
Coca-Cola  160
Cohen, B. C. *et al.*  42, 45
colonialism *see* British colonialism
Communist Party of India (CPI)  19, 130
computers *see* personal computers
concrete  114, 115
Congress Party  129–130, 133
Connell, R. W.  68
Consumer International (CI-ROAP)  9, 146
consumption
  definition  3–5
  and economic liberalisation  132–136, 150–151
  individual, social and material contributions to  5–7
  and religiosity  143–145
  vs consumerism  124–125
cooking appliances  61–62, 138–139
cooling practices *see* air conditioners; house construction and design; refrigerators

cosmetics  40–42, 45–46, 48
  ayurvedic  111
  *bindi*  47
  body odour  110
  fairness (skin-lightening) products  38–40, 153–154, 168
  television advertising  153–154, 163
Counihan, C.  43–44
cultural traditions
  and consumption  135–136
  and social relations  168–170
Curtis, V.  108

Das, K.  57, 58
death rituals  74
Dempsey, C. G.  145
Devi, L. K.  55
Devi, R. P.  140
Devika, J.  15, 22, 58, 71
devotionalism  141–143
dieting  43, 44
disease and cleanliness  107–108
domestic helpers  50, 54, 55, 60, 61, 62, 66
domestic routine  49–53
domestic violence  16
dowry  168
  caste practices  32–33, 79–80
  complaints and retributions  49
  constituents  84–85
  as gift exchange  79, 82, 93
  in household economy and consumption  78–80
  marshalling  86–88
  negotiations  80–84
  recovering cost of education  86
  reforms  79
  and work migration  93, 95
dress *see* clothes
dual residence, work migration  101–103

Eapen, M. and Kodoth, P.  16, 33
economic development  130–132, 136
economic liberalisation  132–136, 150–151
  *see also* globalisation; transnational companies (TNCs)

education 16
  dowry and embedded cost of 86
  Kerala Model of Development 137
  private tutoring 52, 86
  private vs public 139–140
  reforms 58
  women 16, 27, 33, 58–59, 86
  for work migration 92
employment
  Kerala Model of Development 137–138
  women see under women
energy production 175
environmental issues 4–5, 173–176
  global–local relationship 167–168
ethnographic studies 2
extended/joint family 68–73
  education, embedded cost of 86
  events and celebrations 73–78
  and nuclear family relations 49–50, 71–73, 169–170
  support by migrant workers 94, 99–101
  Western vs Keralan 68
  see also dowry
Ezhava caste 24–28
  work migration 91–94

fairness (skin-lightening) products 38–40, 153–154, 168
Falk, P. 159
family
  and consumption 168–169
  reforms 20–22, 70, 71
  see also extended/joint family; marriage
family planning programmes 71
fast food 62–63
feminist vs Hindu perspectives on women 57–58
Fernandes, L. 125, 131–132, 161
*Financial Express* 108
food preparation 61–65
  daily routines 51, 52, 53
  migrant workers 63, 94, 102–103
foreign broadcasting stations 151, 164
foreign cars 122–123

foreign investment 132, 133, 150–151
foreign products 132–135
foreign TV serials 150, 151, 164
freedom
  consumption as expression of 134–135, 146, 171
  of nuclear family 71
  of women 33, 41, 46, 47–48, 66, 168
Freeman, C. 48, 67
Frøystad, K. 2–3
frugality/thrift, intergenerational differences 70, 130, 146–147, 173
Fuller, C. J. 24, 32, 57, 142, 144

Garon, S. 66, 71, 139, 146
gender relations/roles
  differences in mobility 34–35
  reforms 20–22, 31
  sources of ideology 56–60
  task sharing in the home 53–56, 67
  see also men; women
germs and soap 109, 110
Ghandi, Mahatma Mohandras
  dress 37, 129
  'exorcism' of 133–134, 135
  legacy of 128–130, 136
  and Nehru 130–131
  *swadeshi* (indigenous development) 129, 130–131, 132, 139, 147
Ghandi, Rajiv 131–132
Ghandi, Sanjay 123
Gledhill, C. 151, 158
global contact and migration, history of 17–19
global–local relationship 7–8, 15
  environmental issues 167–168
globalisation
  forms of 165
  'opening' to global capitalism 20, 29, 127, 165, 173
  see also economic liberalisation; transnational companies (TNCs)
gold dowries 84, 85, 86–87
Goody, J. and Watt, I. 158

Gough, K.  20, 23, 24, 32, 73, 77
Gulf *see* work migration
Gulf War (Iraqi invasion of Kuwait)  96, 132
Gupta, D.  62
guru worship (devotionalism)  141–143

habitus of everyday consumption  102
hand washing  107, 108, 109
Hansen, K. T.  6, 67
health development  137
herbal products  106, 111
Hindus  22, 23–24
   daily prayers  52–53, 140
   devotionalism  141–142
   family rituals  73–78
   food beliefs  64
   gender roles in the family  56–57
   purity issues  62, 105
   religiosity  140–141
   and consumption  143–144
   *Sabarimala* pilgrimage  75–77, 141, 144
   *Vishu*  75
   work migration  90
   *see also* Ezhava caste; Nair caste
*The Hindu*  121
historic perspective
   family and gender relations reforms  20–22
   global contact and migration  17–19
   Kerala socialism  19–20
   religion and caste system  22–25
Hooper, B.  67
hot water consumption  106–107
house construction and design  112
   alternative  118–119, 175–176
   changes in  113–116, 119–120, 126, 166–167
   migrant workers  93–94, 114–115, 119
house size  114–115
household appliances  51, 55–56
   attitudes to  121
   caste differences in ownership  28
   cooking  61–62, 138–139
   as dowry  84, 85
   men's use of  55
   microwave ovens  65
   migrant workers' ownership  92–93, 94, 101–102
   Gulf vs non-Gulf families  98, *99, 100–101*
   mixies (mixmasters)  51, 56, 62, 148
   refrigerators  28, 56, 63–65, *100*
   television advertising  160–161, 162, 163
   as time-saving devices  66–67
   washing machines  28, 56, 60–61
household income
   of castes  25–26
   and dowry economics  86–87
   work migration  89–90, 92, 93, 103
   Gulf vs non-Gulf families  98, *99, 100*
housewives, modern  49–67
housing density  116

identity issues
   middle class  169
   work migration  97–98
import barriers, removal of  132, 133
income *see* household income
'inconspicuous technologies'  104, 121, 172
independence movement  129–130
*India Today*  43
indigenous development (*swadeshi*)  129, 130–131, 132, 139, 147
infant mortality  107, 108
inheritance law reforms  21–22, 24
intergenerational differences  63, 64–65, 70, 130, 134, 146–147, 173
International Monetary Fund (IMF)  132
internet  13
Isaac, T.  137, 138

Jeffrey, R.  16, 19, 106, 114, 129, 130, 139
Jhally, S.  159
Johnson, M.  47–48, 90, 94
joint family *see* extended/joint family

## Index

joint ventures (TNCs)   132, 133, 150–151
Joseph, A.   33, 34–35

*karnavar* (senior male in matrilineal households)   20–22, 69, 70
Katmandu, Nepal   10, 28–29, 135, 169
Kerala Model of Development   20, 136–138
  and middle class   139
  and 'People's Plan'   138–140
*Kerala Sasatra Sahita Parishat* (KSSP/People's Science Movement)   138
Kerala socialism   19–20, 136–140
Kerala study   10–13
  background to   1–10
Khilnani, S.   130, 132
Kulkarni, D.   143–144

land
  and property as dowry   84–85, 86–88
  reforms   19–20, 23, 24, 25, 84, 87–88
Latour, B.   172
Lee, M.   59
Lever Brothers   110
Liechty, M.   10, 28–29, 35, 135, 145, 160, 164, 169
literacy   16, 107–108, 138
local entrepreneurship   138–139
local government (*Gram Sabhas*)   138
low prices/quality goods   135
*lungi* and men's clothing   37–38, 129
Lury, C. and Warde, A.   132

*makkathayam* (matrilineal inheritance)   20, 21
Mankekar, P.   9, 150, 161, 164, 171
marriage
  constraints, ambiguities and changes   48
  educational achievement and prospects for women   33, 86
  parental choice of partners   32
  *sambandham*   20, 21

  same- and cross-caste   32
  social scrutiny of bridal candidates   33–34
  *see also* dowry
marriage brokers   81
marriage gifts, dowry as   79, 82, 93
Marriot, M. and Inden, R. B.   64
*marumakkathayam* (matrifocal residence and matrilineal inheritance)   20–21, 65, 72
Maruti (cars)   123, 124
Marxism *see* socialism
matrilineal households (*taravad*)   20–22
Mazzarella, W.   47, 130, 133, 134, 136, 160, 166
men
  dowry negotiations   80–83
  dress   37–38, 129
  household economy and consumption decisions   69–70
  *karnavar* (senior male in matrilineal households)   20–22, 69, 70
  *Sabarimala* pilgrimage   75–77, 141, 144
  use of cosmetics   42
  use of household appliances   55
microwave ovens   65
middle class
  decline of frugality   146, 173
  definitions   9–10
  history of growth   20
  identity and consumption   169
  and Kerala Model of Development   139
  leftism and consumerism   140
  religiosity   142–143, 145
  vs local people, environmental consequences of consumption   175
migration
  history of global contact and   17–19
  *see also* work migration
Miller, D.   7, 8, 103, 121, 144–145, 169
Mills, M. B.   8–9, 31, 35
mixies (mixmasters)   51, *56*, 62, 148
mobile phones   34, *99*

mobility, gender differences in 34–35
modernity 172–173
 and tradition 8–9
 'close distance' concept 47, 136, 160
Moore, H. 3, 58
motorcycles 28, 34, 122, 162
Mukhopadhyay, C. C. and Seymore, S. 57–58
Mumbai 95
Murickah, S. J. J. 32, 58
Muslims 22, 23, 84, 90

Nair caste 23–24, 25–28
 domestic routine 49–53
 dowry 32–33, 80–84
 *taravad* (matrilineal households) 20–22
Nambuthiri Brahmin caste 23, 24
Nehru, J. 130–132, 136
'New Age' devotionalism 143
nuclear and extended family relations 49–50, 71–73, 169–170

O'Dougherty, M. 9–10, 124, 169
openness 15, 16
 see also economic liberalisation; globalisation; modernity
Osella, F. and Osella, C. 25, 28, 38, 105, 144

Parayil, G. 137
'People's Plan' 138–140
Pepsi 160
performance see social performance
*perideel* ('naming ceremony') for babies 73
personal computers
 internet 13
 ownership 28, 99, 100
Philippines 47–48, 90, 94
philosophical traditions, Western vs Indian 143–144
Phookan, M. 38, 41, 42, 111
pilgrimages
 Christian 144
 Hindu *Sabarimala* 75–77, 141, 144

political ideologies see Ghandi, Mahatma Mohandras; Kerala socialism; globalisation; economic liberalisation
practice-theory approach 2, 6
pregnancy, first 77–78
property
 and land as dowry 84–85, 86–88
 law reforms 21–22, 24
pyramid selling ('Tupperware Model') 111

Rai, Aishwarya 42, 43, 160
refrigerators 28, 56, 63–65, 100
religion
 and caste relationship 22–23
 see also Christians; Hindus; Muslims
religious holidays 74–75
religious reformers 21
rituals
 cleansing 105
 family 73–78
Rowchowdry, A. 132

*Sabarimala* pilgrimage 75–77, 141, 144
*sambandham* (marriage) 20, 21
Saradamoni, K. 95
savings rates 146
Seabrook, J. 132
self-help groups 138–139
self-reliance (*swara*) 131
Sen, Sushmita 42, 43
Shankar, Shri Ravi 143
Shiva, V. 109
Shove, E. 104, 121, 126, 172
Simpson, B. 88
skin-lightening (fairness) products 38–40, 153–154, 168
Slater, D. 5
soap 107–111
 and cleaning agents 104, 106, 126, 163
'soap operas' 151, 152–158
social performance
 air conditioners 121
 cars 122–123, 127
 consumption as 172–173

social relations and cultural practices 168–170
social scrutiny of bridal candidates 33–34
Social Shaping of Technology (SST) 172
socialism
  gender relations reform 22
  Kerala 19–20, 136–140
socially correct appearance and behaviour (*swabhaavam*) 33–34, 35–36, 48
Soman, C. K. 107, 108
sponsorship 45–46, 150
*Sreejanmam* (Women's Lives) 156–158
Srinivas, M. N. 33
*Sthree Malayalam* (Malayalee women) 152–155
*Sthree oru Jwala* (Woman, Aflame) 155–156
stress
  of family separation, work migration 96–97, 99
*stridhanam* (women's wealth) 32–33
Surendran, P. 137
SUVs (sports utility vehicles) 85, 122–123
*swabhaavam* (socially correct appearance and behaviour) 33–34, 35–36, 48
*swadeshi* (indigenous development) 129, 130–131, 132, 139, 147
*Swadeshi* Programme 139
*swaraj* (self reliance) 131

*taravad* (matrilineal households) 20–22
technology 170–172
  'inconspicuous technologies' 104, 121, 172
  transfer 167–168
television 28
  advent in India and Kerala 150–152
  advertising 158–163
    child-directed 162–163
    development of 150
    impact on consumption 162–163, 164, 166
    products 153–154, 155, 157
  in everyday life 149
  foreign serials 150, 151, 164
  local programmes 149–150, 151, 152–158
  *Sabarimala* pilgrimage 141
  serials ('soap operas') 151, 152–158
  social pressure to conform 147
  social themes 159, 163–164
  theories of effects 149
theory–practice relationship 2–3
transnational companies (TNCs) 132, 133, 134, 136
  and agents 165–168
  television advertising 159–160

UN Human Development Index 138
Unilever 38, 108, 111
USA
  air conditioning 120–121
  car and ideal family 125
  climate change 174
  *see also entries beginning* Western *and* foreign
Usha, V. T. 57, 151, 157–158, 161, 163

Varma, P. K. 130, 131, 133–134, 140, 163
Vinikas, V. 109
*Vishukkani* 75
*Viswakamos* artisan caste 113

Waldrop, A. K. 144
Warde, A. 6, 104, 172
  Lury, C. and 132
washing machines 28, 56, 60–61
water 4–5, 175
  hot water consumption 106–107
  and soap use 108–109
Western beauty ideals 42–48
Western cleanliness ideals 109–110
Western consumption and 'energy efficiency' 174
Western development paradigm vs *swadeshi* 130–131

Western dress   35–36, 37
Western family relationships   68
Western house construction and design   113–114
Western materialism   18, 129
Western New Age Movement   143
Western reformist influences   21, 58
Western trade contacts, history of   17–18
Western vs Indian philosophical traditions   143–144
Wilhite, H.   4, 138–139, 168
  *et al.*   4, 106
  and Lutzenhiser, L.   125
Wilk, R.   7–8, 46, 48, 102, 111, 157, 171
women
  dress   35–37, *46*
  educational achievement   16, 27, 33, 58–59, 86
  employment
    and household roles   58–60, 61, 63, 67
    and ownership of household appliances   55–56, 66–67
  feminist vs Hindu perspectives   57–58
  freedom   33, 41, 46, 47–48, 66, 168
  modern housewives   49–67
  *stridhanam* (wealth)   32–33
  suicide rate   16, 59
  and television
    advertising depictions   161
    serial themes   151, 152–158
    viewing   148, 149, 151
  work migration   47–48
  *see also* beauty; cosmetics; dowry; extended/joint family; gender relations/roles; marriage
work migration   89–91
  air conditioners   *101*, 118
  consumption in 'Gulf' families   98–99, 103
  dowries   84, 85
  dual residence   101–103
  expenses   95, 96
  family issues   99–101
  food 'wastage' ideas   63
  house construction and design   93–94, 114–115, 119
  identity issues   97–98
  impact of Iraqi invasion of Kuwait   96
  local criticism/stereotyping   94–95, 97, 103
  normalising consumption   173
  and *Sabarimala* pilgrimage   76
  stress of family separation   96–97, 99
  unsuccessful attempts   95–96
  whole family   91–94
  women   47–48
World Bank   107, 108, 109, 167
World Systems Theory   7

Zachariah, K. C.
  *et al.*   2, 8, 90, 94, 95
  and Kannan, K. P.   89, 90
  and Rajan, S. I.   89, 91, 98